THE LEGO® POWER FUNCTIONS IDEA BOOK
CARS AND CONTRAPTIONS

VOLUME 2

THE LEGO® POWER FUNCTIONS IDEA BOOK

CARS AND CONTRAPTIONS

YOSHIHITO ISOGAWA

no starch
press

The LEGO® Power Functions Idea Book, Volume 2: Cars and Contraptions.

Copyright © 2016 by Yoshihito Isogawa.

Printed in China
First Printing

19 18 17 16 15   1 2 3 4 5 6 7 8 9

ISBN-10: 1-59327-689-3
ISBN-13: 978-1-59327-689-8

Publisher: William Pollock
Production Editor: Riley Hoffman
Cover Design: Beth Middleworth
Photographer: Yoshihito Isogawa
Author Photo: Sumiko Hirano
Developmental Editor: Tyler Ortman
Technical Reviewer: Sumiko Hirano
Proofreader: Fleming Editorial Services

For information on distribution, translations, or bulk sales, please contact No Starch Press, Inc. directly:
No Starch Press, Inc.
245 8th Street, San Francisco, CA 94103
phone: 415.863.9900; info@nostarch.com
www.nostarch.com

The Library of Congress has cataloged the first volume as follows:

Isogawa, Yoshihito, 1962-
   The LEGO power functions idea book / by Yoshihito Isogawa.
      pages cm
   Summary: "A compilation of small projects to build with LEGO Technic parts, including gears, motors, gadgets, and other moving elements. Contains step-by-step building instructions for rack-and-pinion steering systems, sliding doors, grasping claws, and ball-shooting devices. Explores principles of simple machines, gearing, and power translation"-- Provided by publisher.
   ISBN 978-1-59327-688-1 -- ISBN 1-59327-688-5
   1. Machinery--Models. 2. Power (Mechanics) 3. LEGO toys. I. Title.
   TJ248.I863 2016
   621.8--dc23
                        2015021881

Production Date: 6/24/2015
Plant & Location: Printed by Everbest Printing (Guangzhou, China), Co. Ltd
Job / Batch #: 54666-0 / 708308.1

Contents

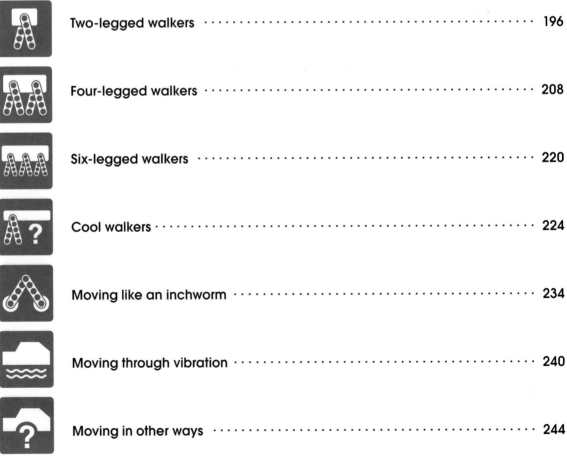
PART 2 • Moving Without Tires

PART 3 • Special Mechanisms

Introduction

This is an idea book, offering hundreds of projects and mechanisms you can build with LEGO Technic parts. The book especially focuses on LEGO Power Functions, which is the latest version of the Technic system of motors, lights, and other electric building elements.

Where Are the Words?

Other than this brief introduction and the table of contents, this book has almost no words. Instead, you'll find a series of photographs of increasingly complex models, each designed to demonstrate a mechanical principle or building technique.

While the book lists the pieces needed to build each model, it does not include step-by-step building instructions. Look at the photographs taken from various angles and try to reproduce the model. Building in this way is something like putting together a puzzle. You'll get the hang of it after a little practice.

The Use of Color

The examples in this book are made with parts of various colors to make it easier for you to see the individual bricks' shapes. But you don't need to use the colors I've chosen in your models; use whichever colors you want to make the projects your own.

Substituting Parts

The parts used in this book were selected from among the easily obtainable ones as much as possible, but you'll probably still be missing a few.

Try to build as many models as possible using the parts that you already own. If you find that you're missing parts, try to think of ways to substitute other parts for the ones that you're missing.

For example, there are many types of LEGO tires. If you don't have the tires shown in a particular project, try using any tires you have that are a similar size. Also, there are several types of Power Functions motors. In this book, the commonly available Medium (M) motors are used most frequently. It's relatively easy to replace the M motor with the Large (L) motor or older motors from earlier systems, so use whatever you have available.

The Parts List in the back of the book will help you find the pieces you need.

You Are the Creator

Look at the models you build closely. By thinking about how they move and why they are designed that way, you will greatly improve your building skills.

This is an idea book; it's about imagination. It is my sincere hope that you make these projects your own, combine them, and evolve them into something even better—your own original models.

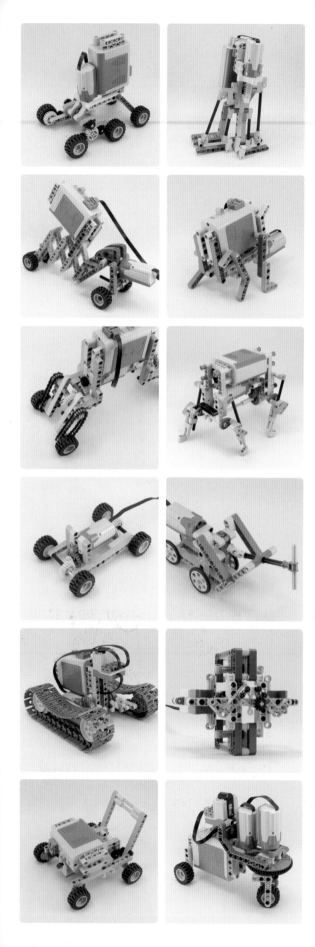

PART 1

Vehicles

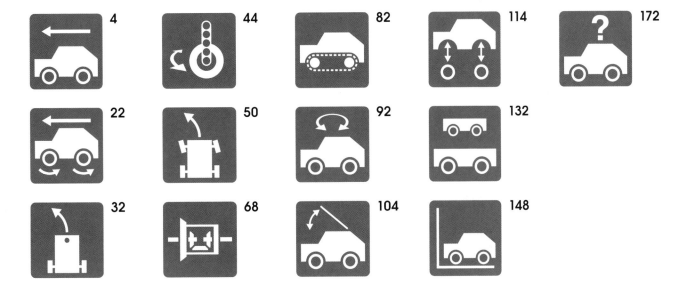

Driving wheels with a motor

#1

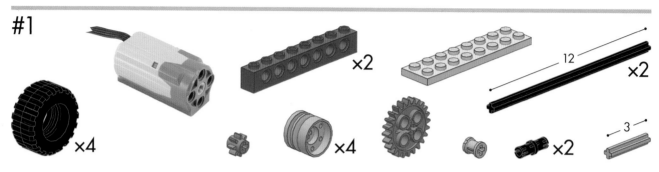

×4   ×2   12   ×2   ×4   ×2   3

8:24 = 1:3

#2

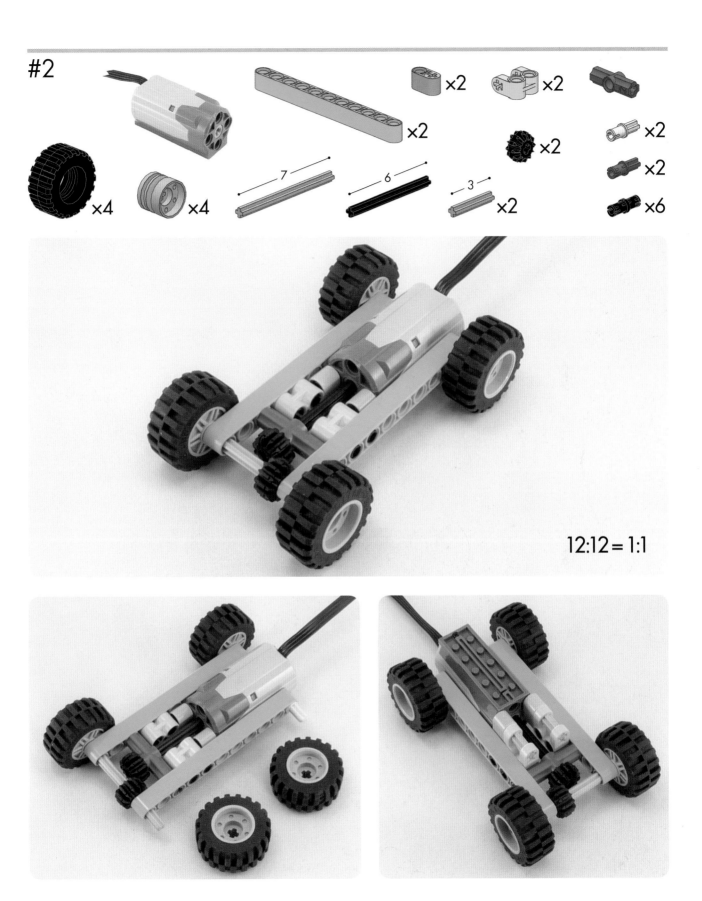

×4 ×4 ×2 ×7 ×6 ×3 ×2 ×2 ×2 ×2 ×2 ×2 ×2 ×6

12:12 = 1:1

#3

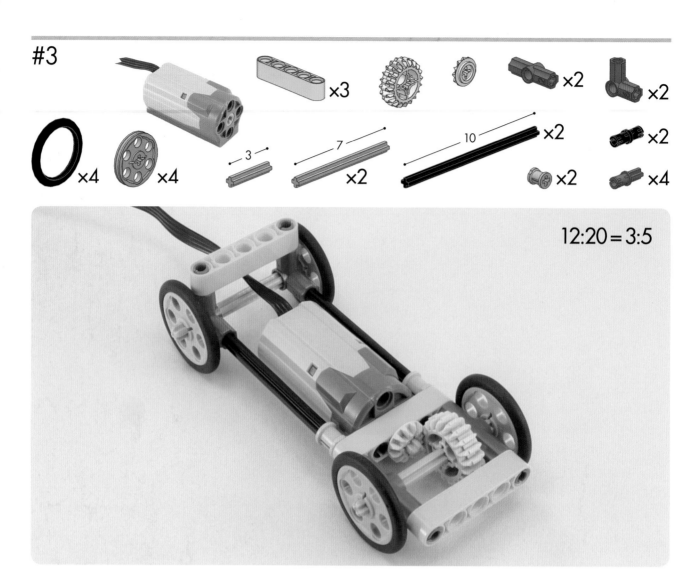

×3 ×2 ×2

×4 ×4 3 7 10 ×2 ×2 ×2 ×4

$12:20 = 3:5$

#4

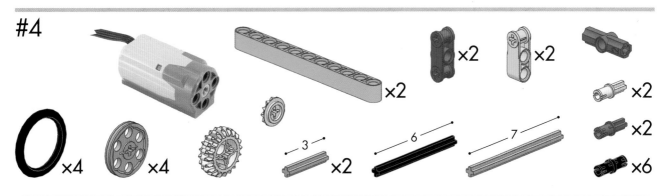

×2 ×2
×4 ×4
3 ×2 6 7 ×2 ×2 ×6

12:20 = 3:5

#5

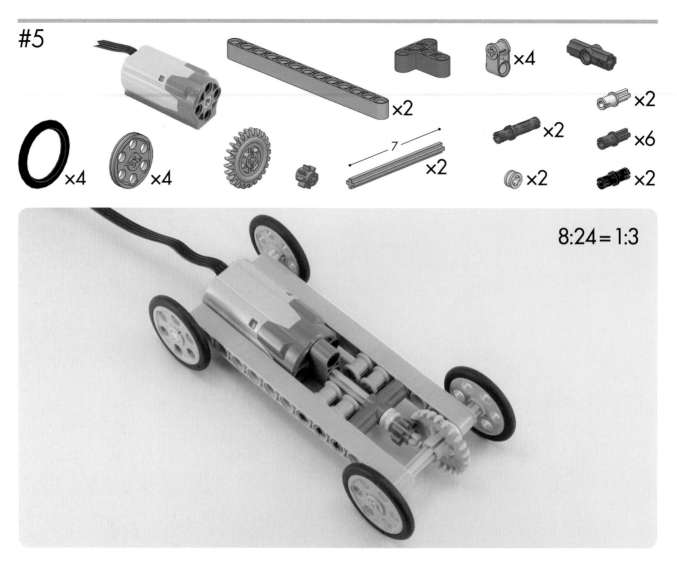

×2

×4

×2

×4

×4

7

×2

×2

×6

×2

×2

8:24 = 1:3

#6

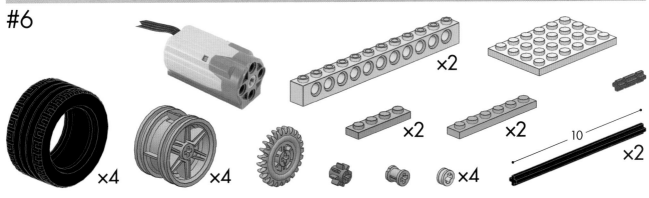

×4 ×4 ×2 ×2 ×2 10 ×2 ×4

8:24 = 1:3

#7

×4 ×4

×2

×3

×2

10

×2 ×4 ×2

×2

×6

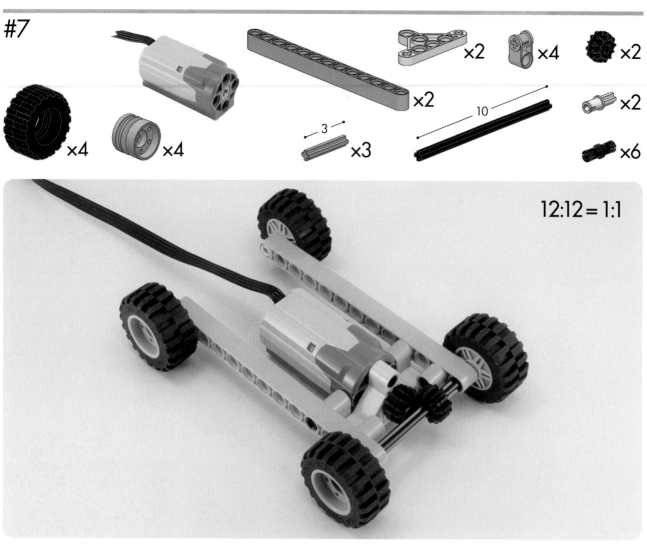

12:12 = 1:1

#8

×4 ×4 ×2

×4 ×5 10 ×2

20:12 = 5:3

#9

×2 ×2 ×2 ×2 ×2 ×2 ×4 8

$20{:}12 = 5{:}3$

#10

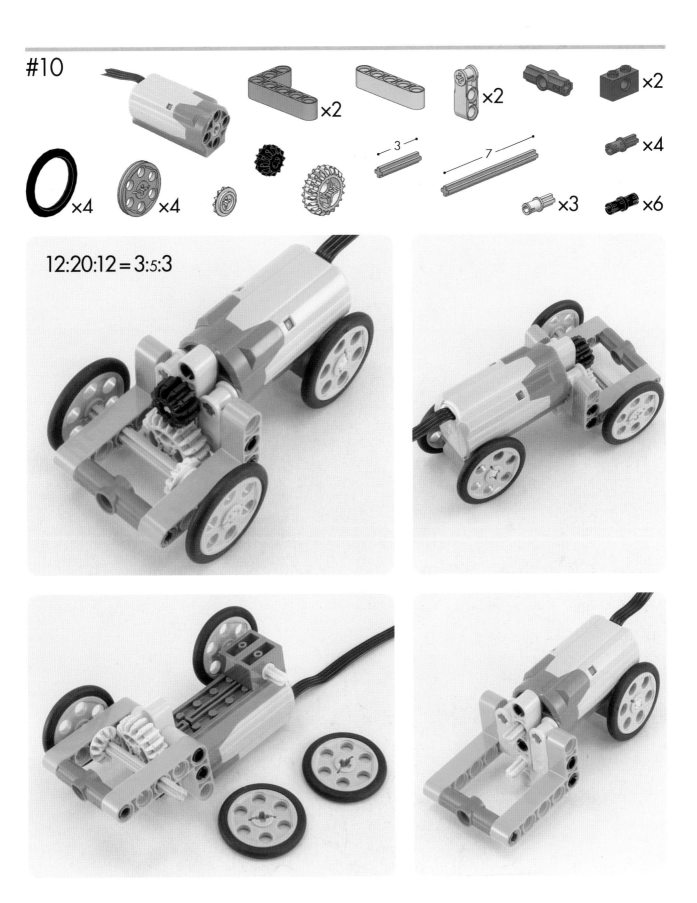

×2 ×2 ×2

×4 ×4 ×4

3

7

×4

×3 ×6

12:20:12 = 3:5:3

#11

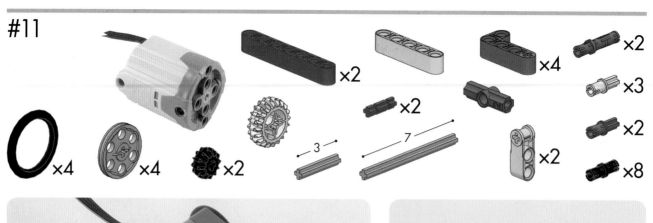

×4 ×4 ×2 — 3 — 7 ×2 ×2 ×2 ×2 ×3 ×2 ×8

20:12:12 = 5:3:3

#12

×2   ×5   ×2

×4   ×4   ×14   3

$$\frac{8:24}{12:20} = \frac{1:3}{3:5}$$

#13

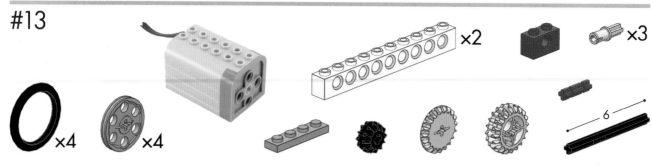

×2 ×3 ×4 ×4 6

20:12:20 = 5:3:5

#14

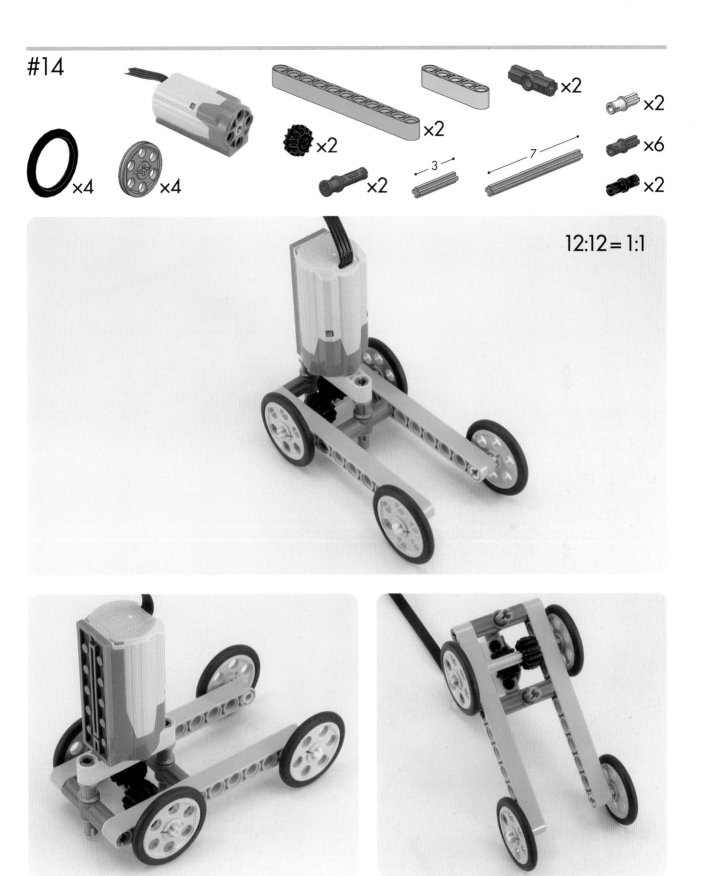

×4 ×4 ×2 ×2 ×2 ×2 ×2 ×6 ×2

12:12 = 1:1

3 7

#15

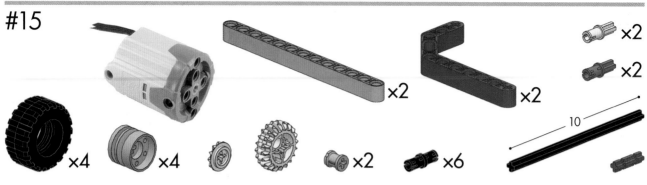

×2 ×2 ×4 ×4 ×2 ×6 ×2 10

12:20 = 3:5

#16

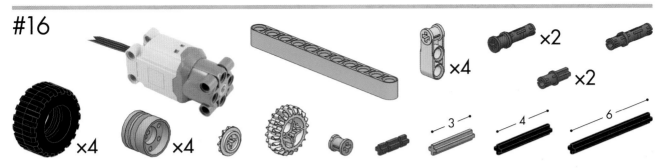

×4 ×4 ×4 ×2 ×2

—3— —4— —6—

12:20 = 3:5

#17

×2 ×2 ×2

5 ×2 6 8

×4 ×4 ×2 ×3 ×6

1:8

#18

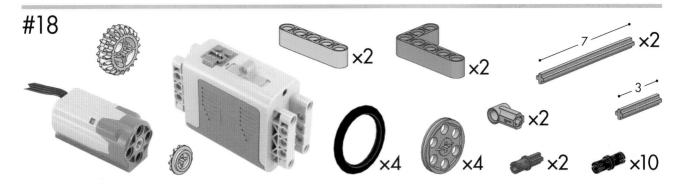

×2 ×2 7 ×2 3 ×4 ×4 ×2 ×2 ×10

12:20 = 3:5

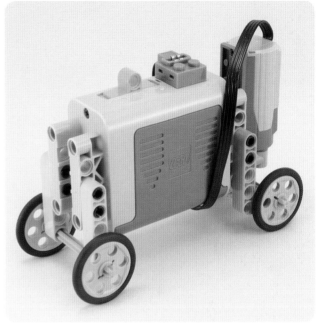

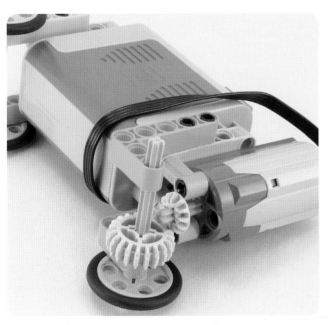

4WD cars

#19

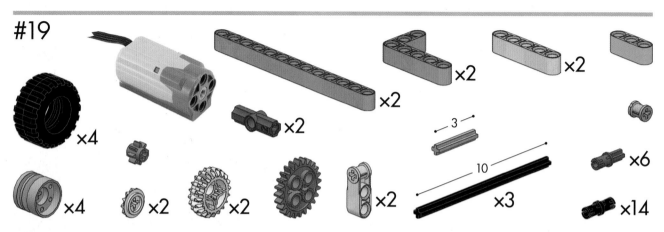

×4 ×2 ×2 ×2 ×2

×4 ×2 ×2 ×2 3 10 ×3 ×6 ×14

#20

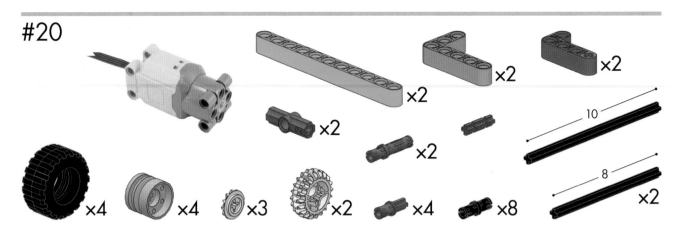

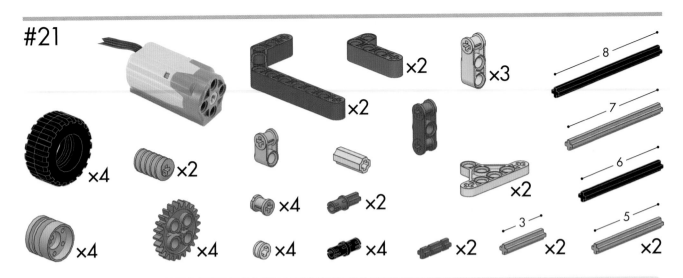

×4 ×2

×4 ×4 ×4

×2 ×4 ×2

×2 ×3

×2

×4 ×2

8
7
6
3 ×2 5 ×2

#22

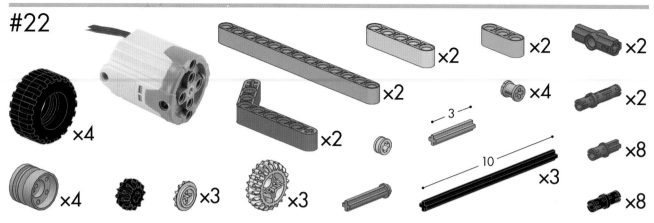

×4 ×2 ×2 ×2 ×2 ×2

×2 ×4 ×2

3

×4 ×3 ×3 10 ×3 ×8

×8

#23

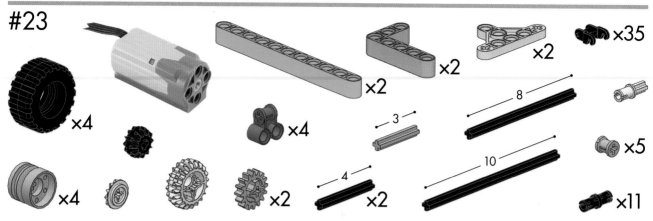

×4 ×4 ×2 ×2 ×35

×4 ×4 ×2 3 4 ×2 8 10 ×5 ×11

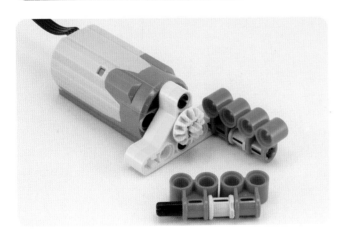

You can build cars that steer by combining these with the models in **Caster wheels** (page 44).

Each motor turns a wheel

#24

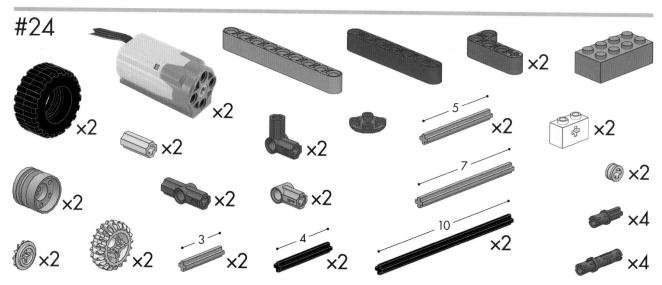

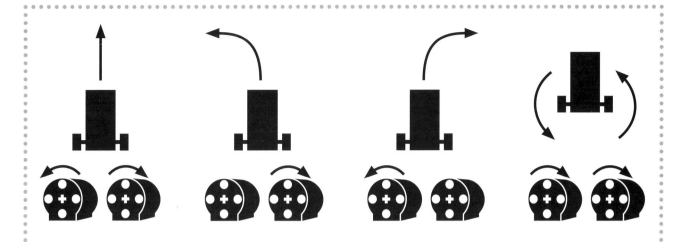

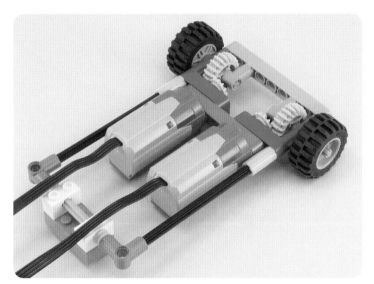

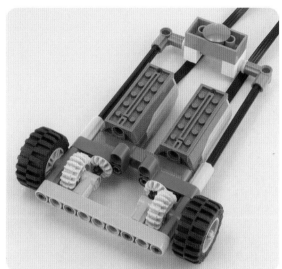

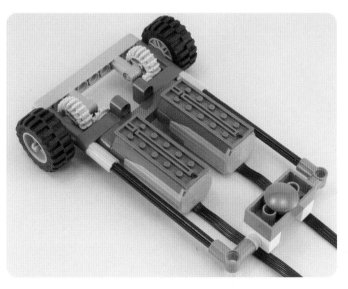

#25

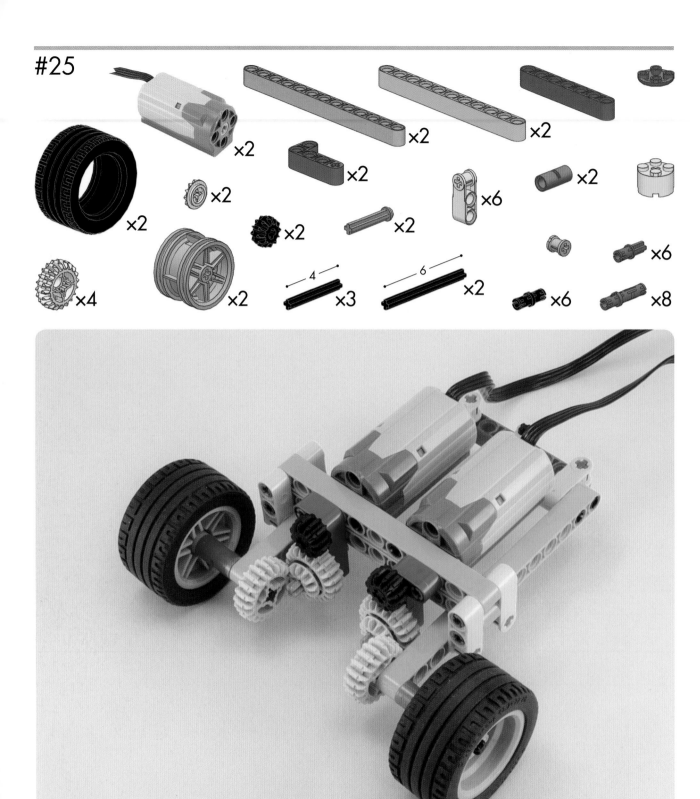

×2 ×2 ×2 ×2 ×2

×2 ×2 ×6 ×2

×2 ×2 ×2 ×6

×4 ×2 4 ×3 6 ×2 ×6 ×8

Each motor turns a wheel

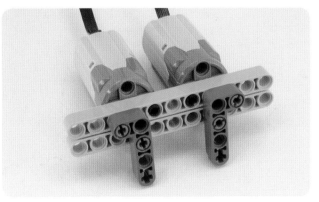

#26

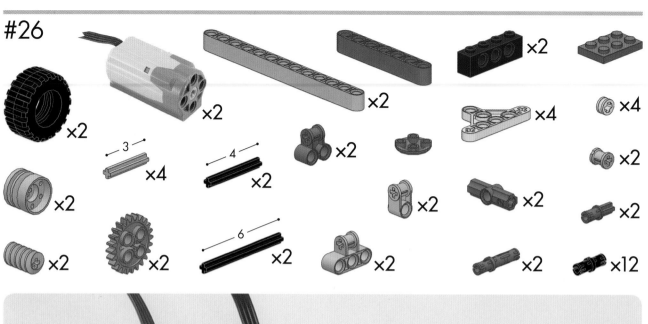

#27

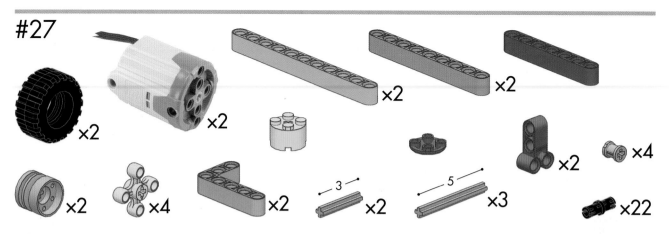

×2 ×2 ×2 ×2 ×2

×2 ×4 ×2 3 ×2 5 ×3 ×2 ×4 ×22

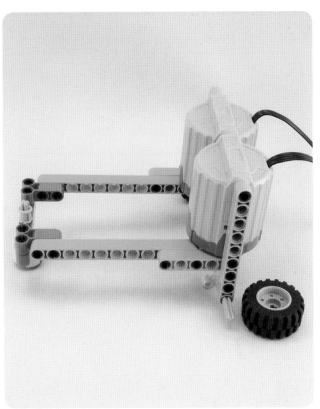

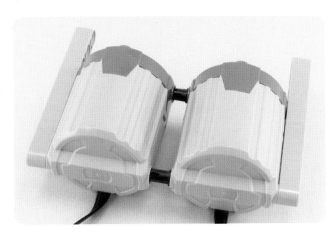

#28

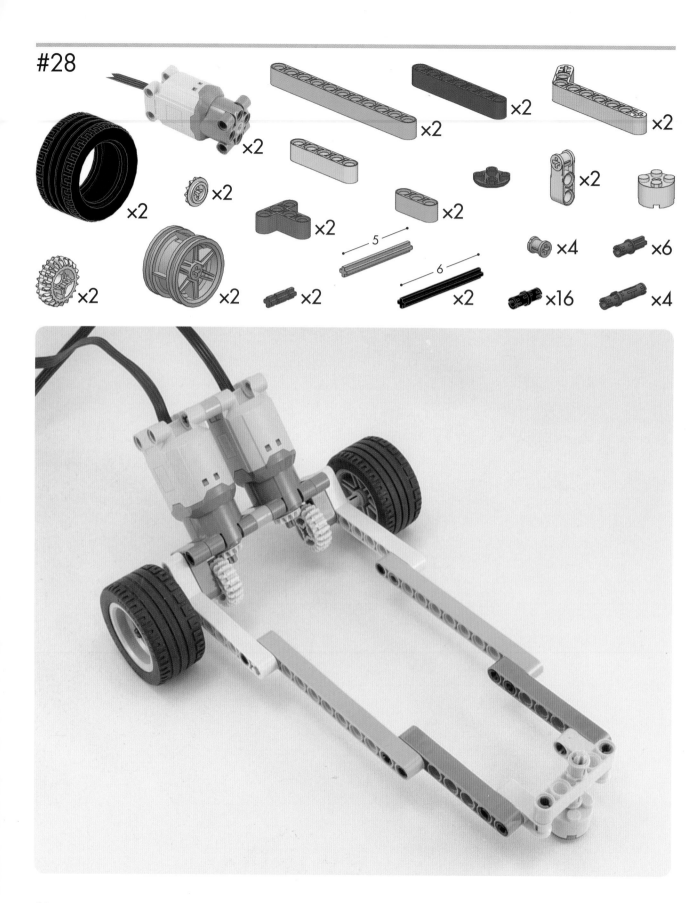

×2 ×2 ×2 ×2

×2 ×2 ×2 ×2

×2 ×2 ×2 ×2

×2 ×2 5 6 ×4 ×6

×2 ×2 ×16 ×4

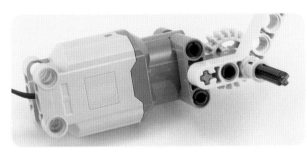

#29

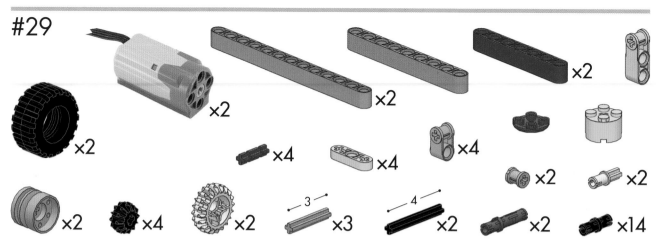

Each motor turns a wheel

You can build cars that steer by combining these with the models in **Each motor turns a wheel** (page 32).

Caster wheels

#30

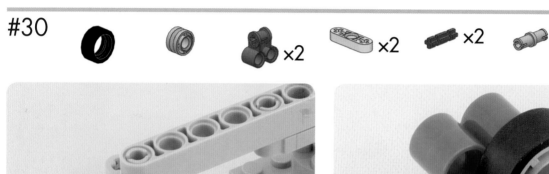

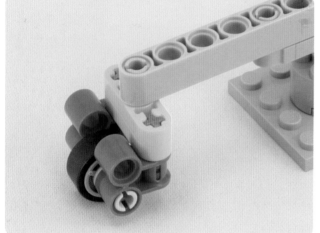

#31

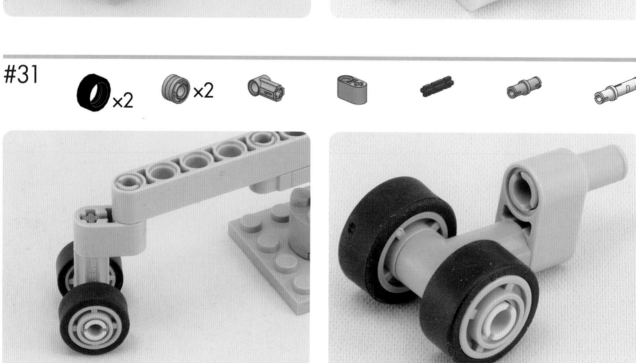

#32

×2 ×2

#33

×2 ×3

#34

 ×2 ×2 ×2 ×2 3  ×2 ×2

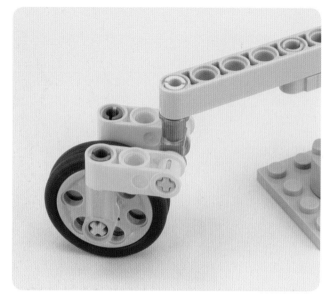

#35

 ×2 ×2  ×2   ×2   ×2

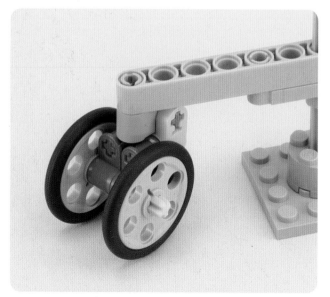

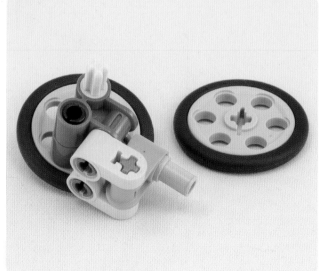

x

#36

$\times 2$ $\times 2$ $\times 2$ $\times 2$ $\times 2$ 3 $\times 2$ $\times 4$

#37

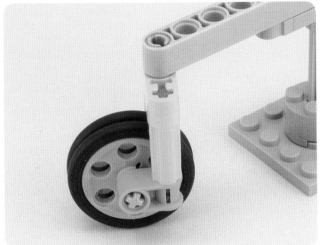

$\times 2$ $\times 2$ $\times 2$ $\times 2$ $\times 2$ $\times 2$ $\times 4$ $\times 2$

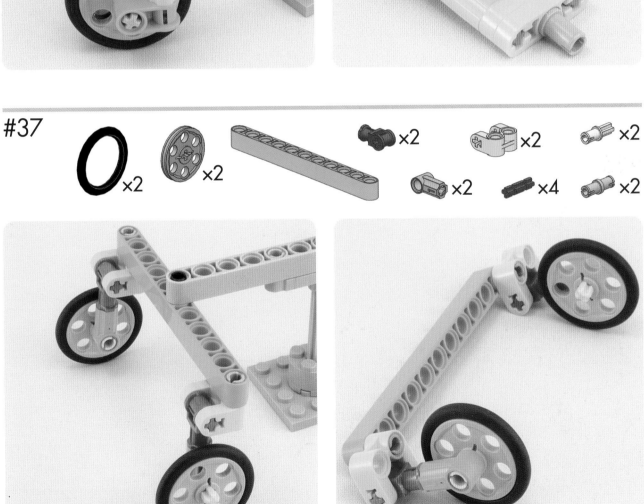

#38

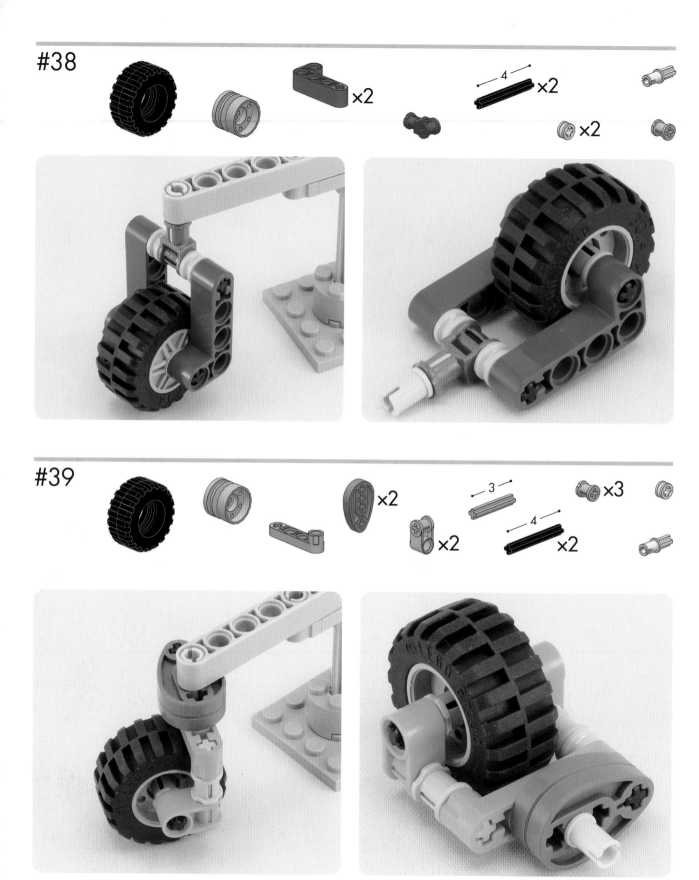

#39

#40

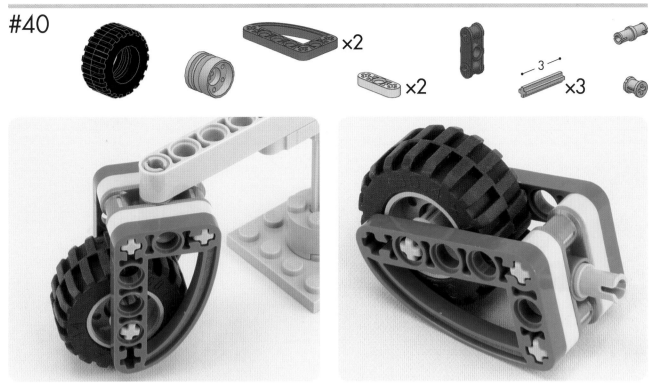

×2

×2

3

×3

#41

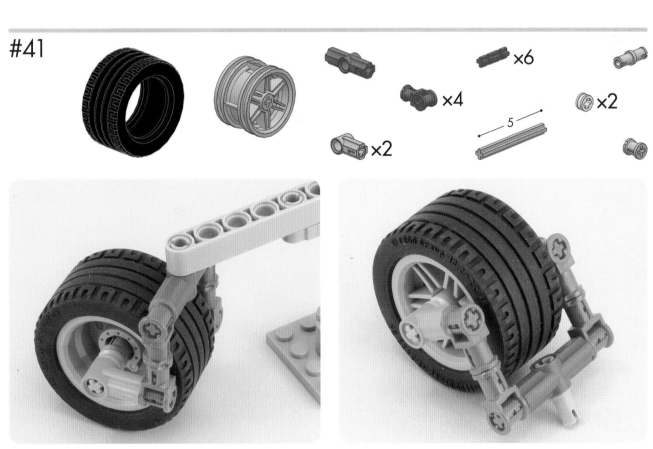

×6

×4

×2

5

×2

You can combine these models with those in **Differential gears** (page 68).

Steering with the servo motor

#42

×4

×4

×4

×2

×2

×2

×2

×4

3
×4

4

8

×12

×2

×2

×4

×2

×6

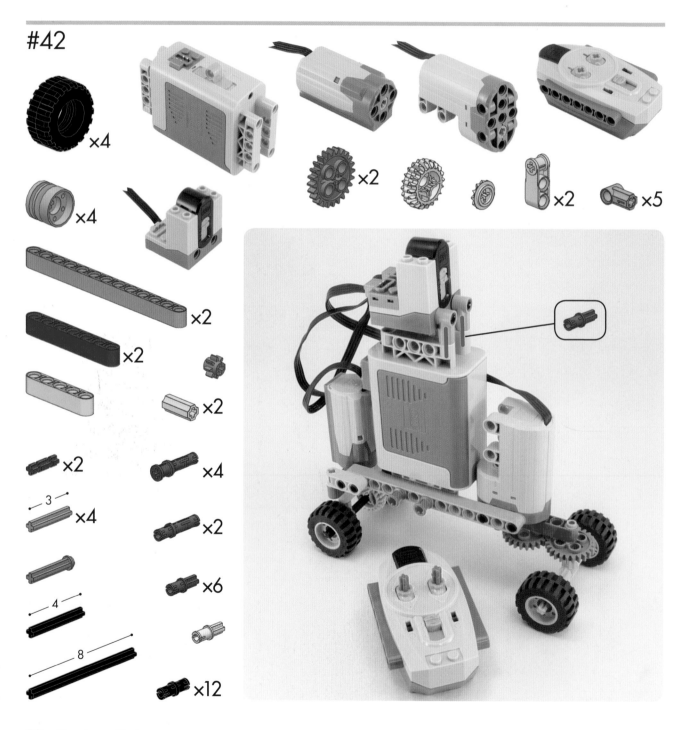

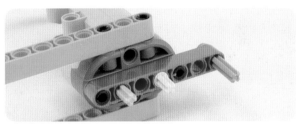

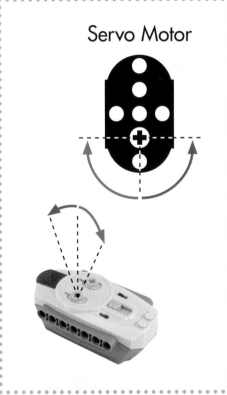

Servo Motor

#43

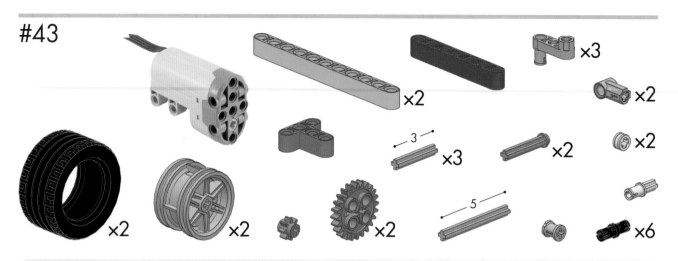

×2 ×3 ×2 ×2 ×2 ×3 ×2 ×2 ×2 ×5 ×6

#44

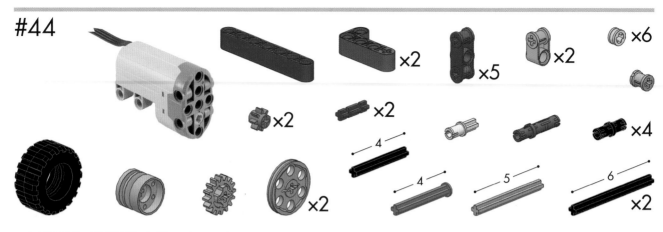

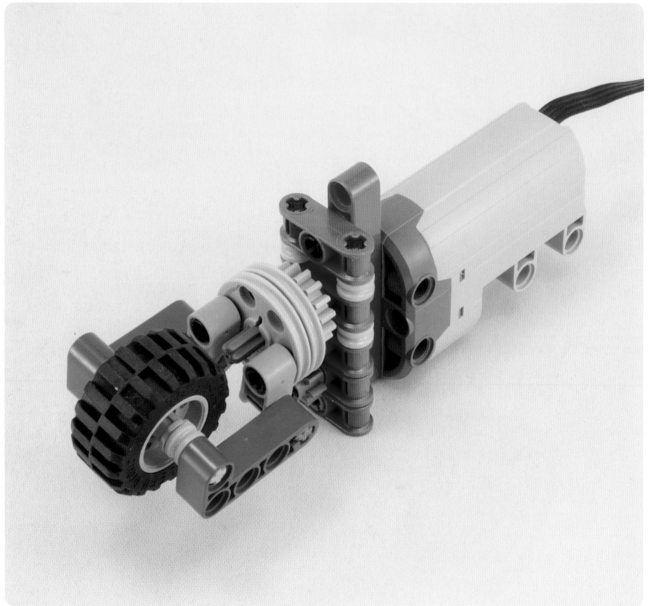

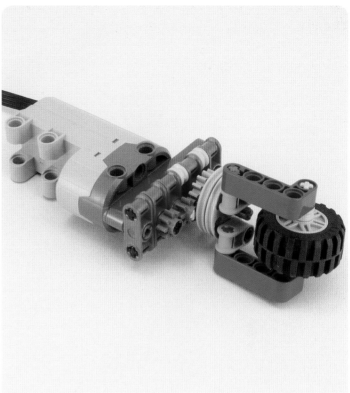

#45

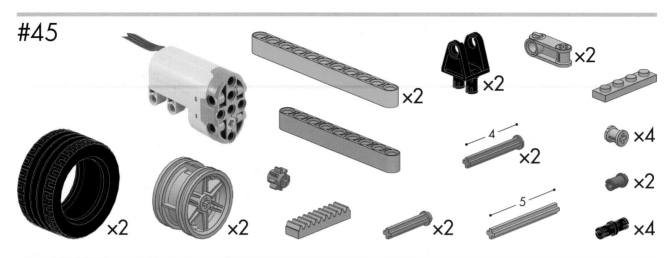

#46

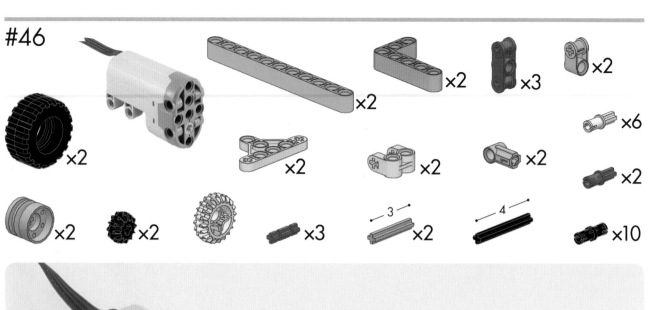

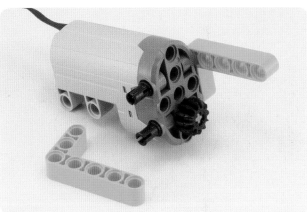

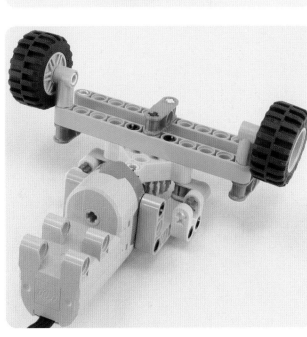

#47

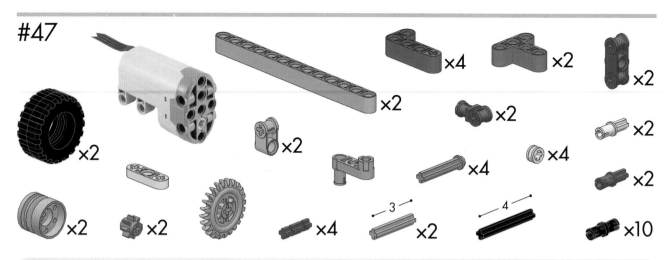

×2 ×2 ×4 ×2 ×2 ×2 ×2 ×2 ×2 ×4 ×4 ×4 ×2 ×10

#48

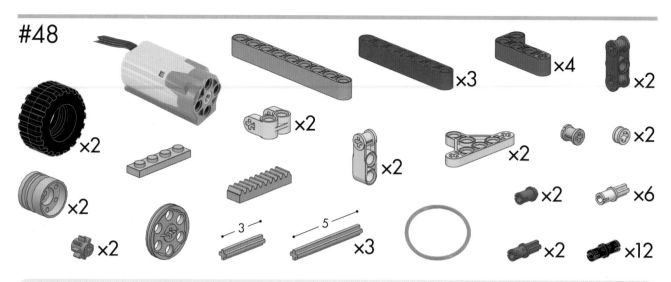

×2 ×2 ×3 ×4 ×2

×2 ×2 ×2 ×2

×2 ×2 ×2 ×6

×2 3 5 ×3 ×2 ×12

#49

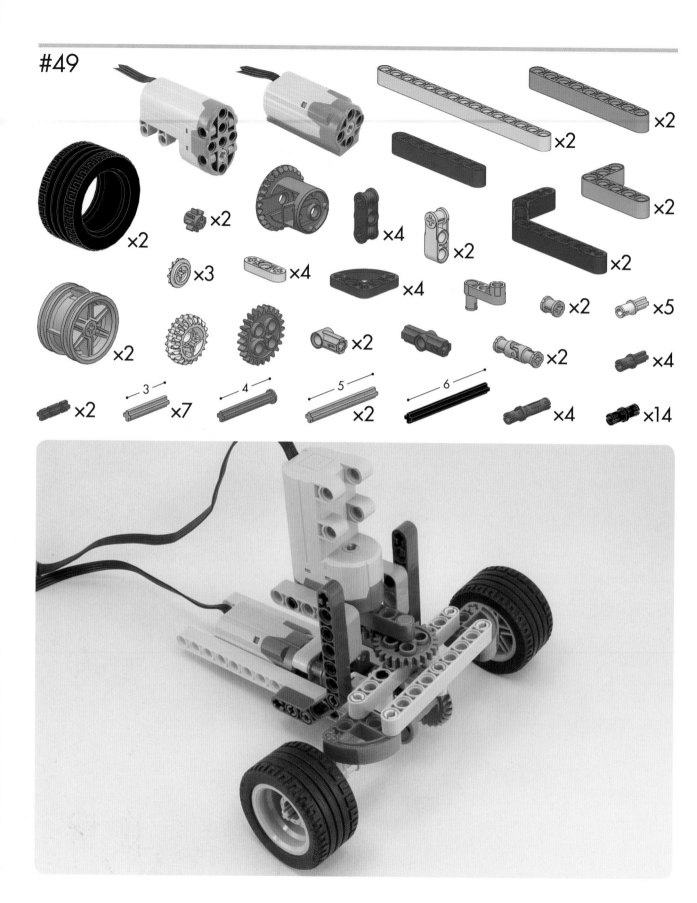

×2 ×2 ×2 ×2 ×4 ×2 ×2 ×2 ×2 ×3 ×4 ×4 ×2 ×5 ×2 ×2 ×4 3 ×2 ×7 4 5 ×2 6 ×4 ×14

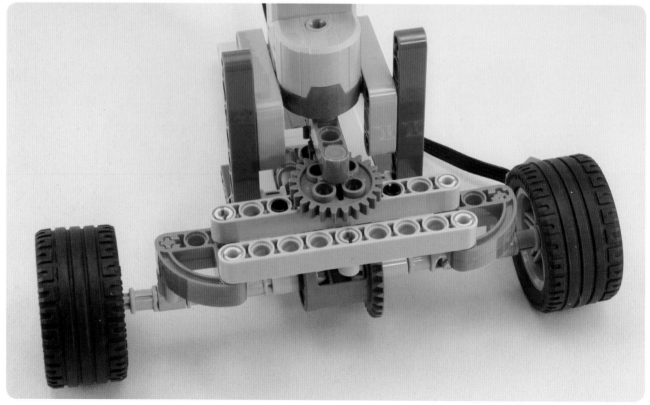

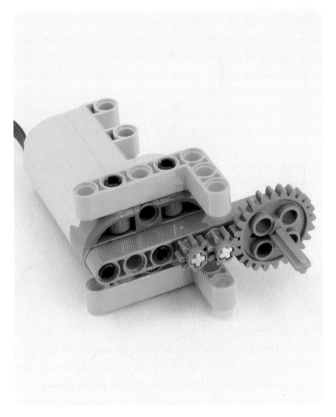

You can combine these models with those in **Steering with the servo motor** (page 50).

Differential gears

#50

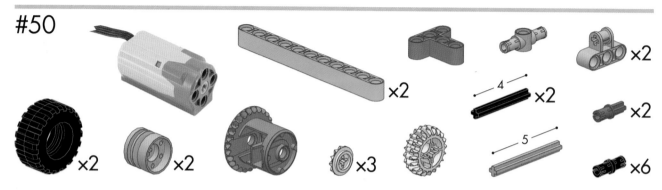

×2 4 ×2 ×2

×2 ×2 ×3 5 ×6

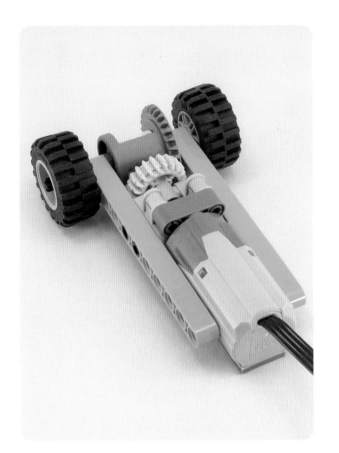

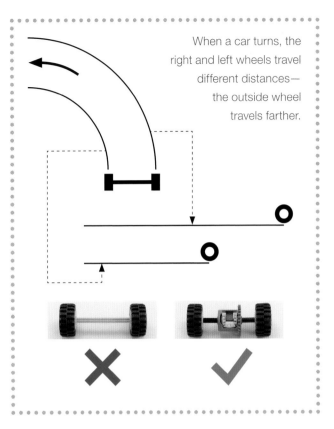

When a car turns, the right and left wheels travel different distances—the outside wheel travels farther.

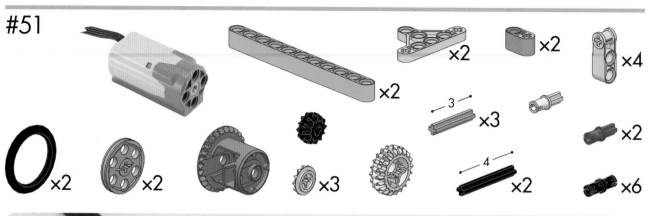

×2 ×2 ×2 ×4

×3 ×3 ×2

×2 ×2 ×3 ×2 ×6

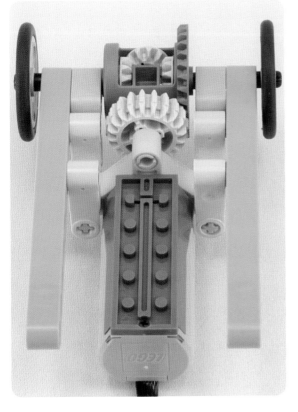

#52

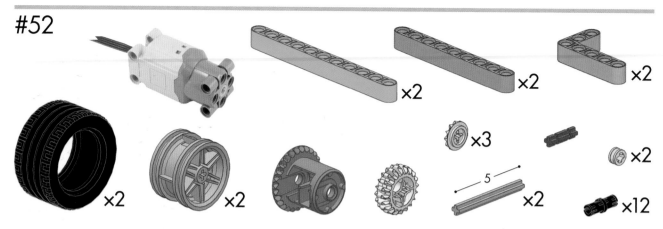

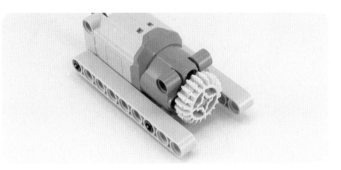

#53

×2 ×3 ×2 ×2 ×2 ×2 ×4

4

5

#54

×2
×2
×2
×2
×2
×2
×2
×2
×3
×3

5
6

#55

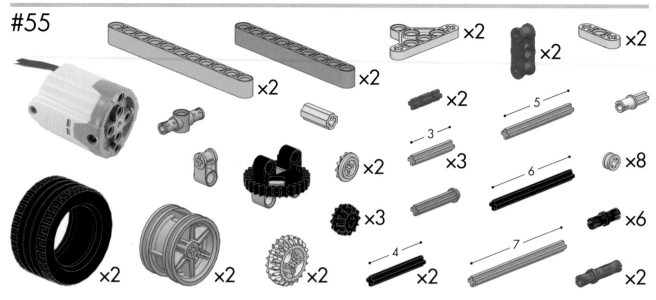

×2 ×2 ×2 ×2 ×2 ×2 ×3 ×3 ×8 ×2 ×3 ×2 ×2 ×6 ×2

3 4 5 6 7

#56

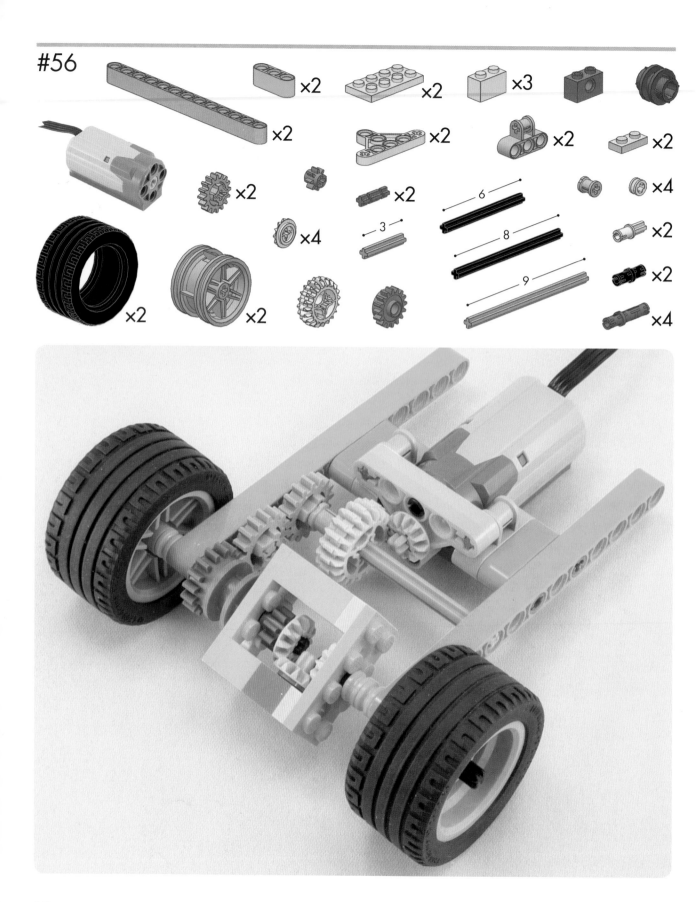

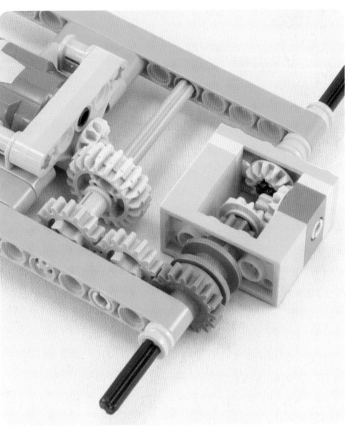

Crawlers

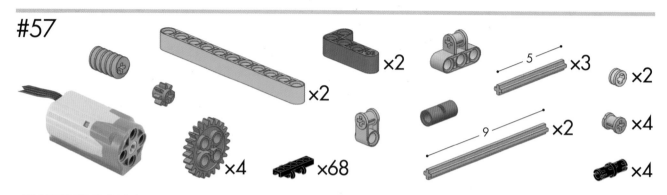

×2 ×2 ×3 ×2

×4 ×68 ×2 ×4 ×4

#58

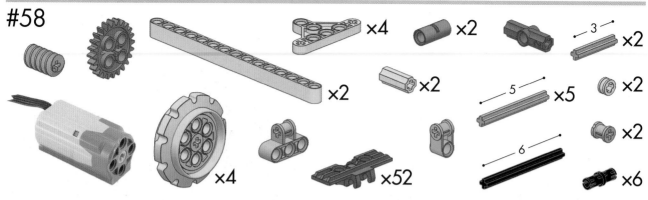

×4 ×2 ×2 ×2 ×2 ×2 ×5 ×4 ×52 ×6

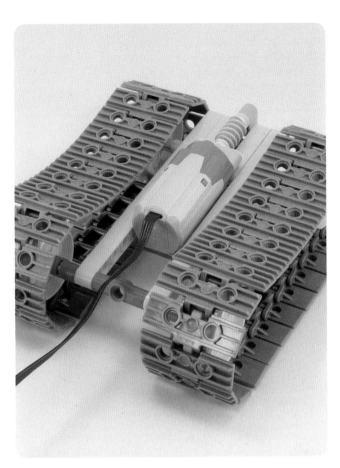

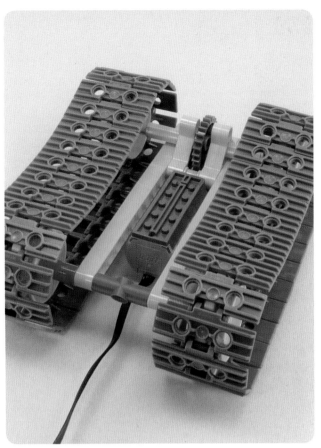

#59

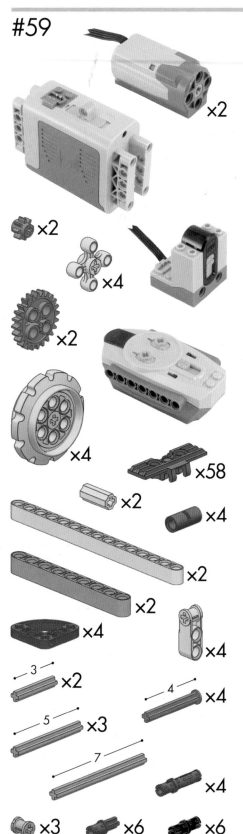

×2

×2

×4

×2

×4

×58

×2

×4

×2

×2

×4

×4

3 ×2

5 ×3

7

4 ×4

×4

×3 ×6 ×6

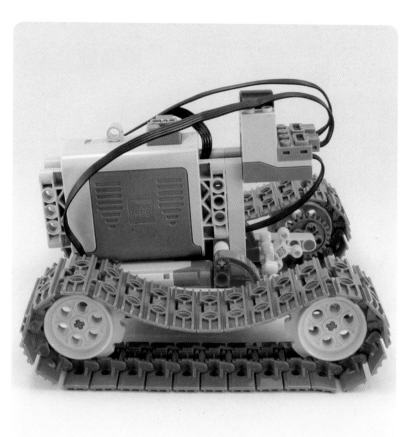

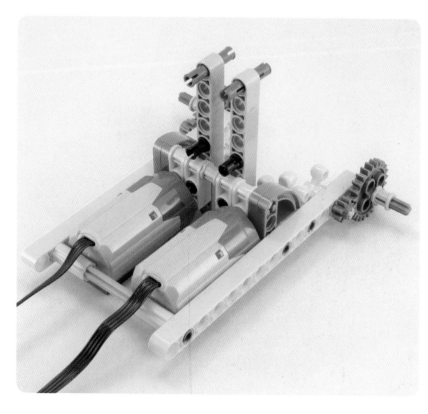

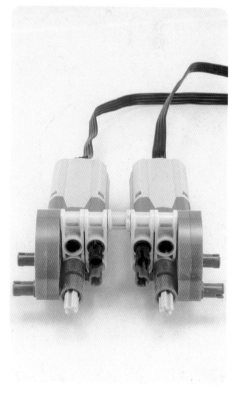

#60

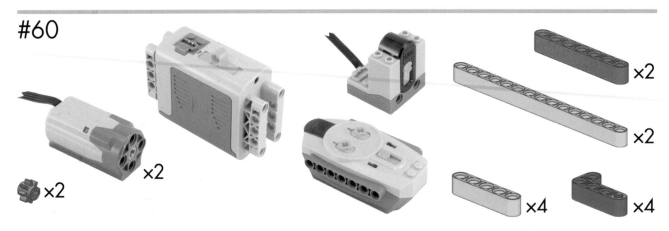

 ×4

 ×8

 ×98

 ×2

 ×8

 ×20

 ×10

#61

×2 ×2 ×4 ×2

×2 ×3 ×2 ×2 ×54

×5 ×2 ×2 ×2

×2

3 ×4

4

5 ×3

×2

×4

×3

×6

×2

×4

×24

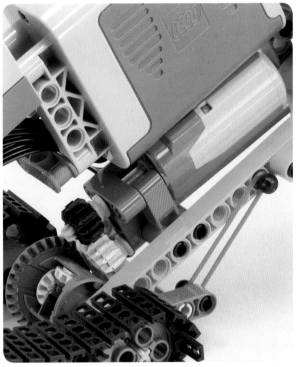

Cars that spin something

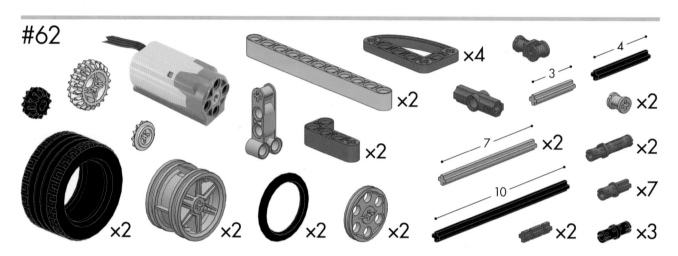

×4

×2

3

×2

4

×2

×2

×7

7

×2

×2

10

×2

×3

×2

×2

×2

×2

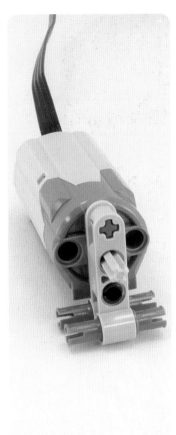

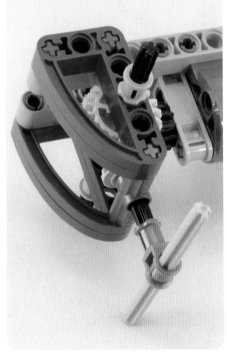

#63

×4 ×4 ×2 ×4 ×2

×2
4 ×2
4
5
6
8
×2
×4

#64

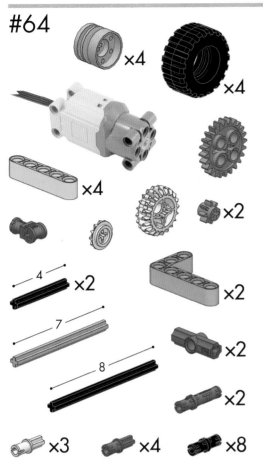

×4
×4
×4
×2
×2
×2
×2
×3 ×4 ×8

4
7
8

#65

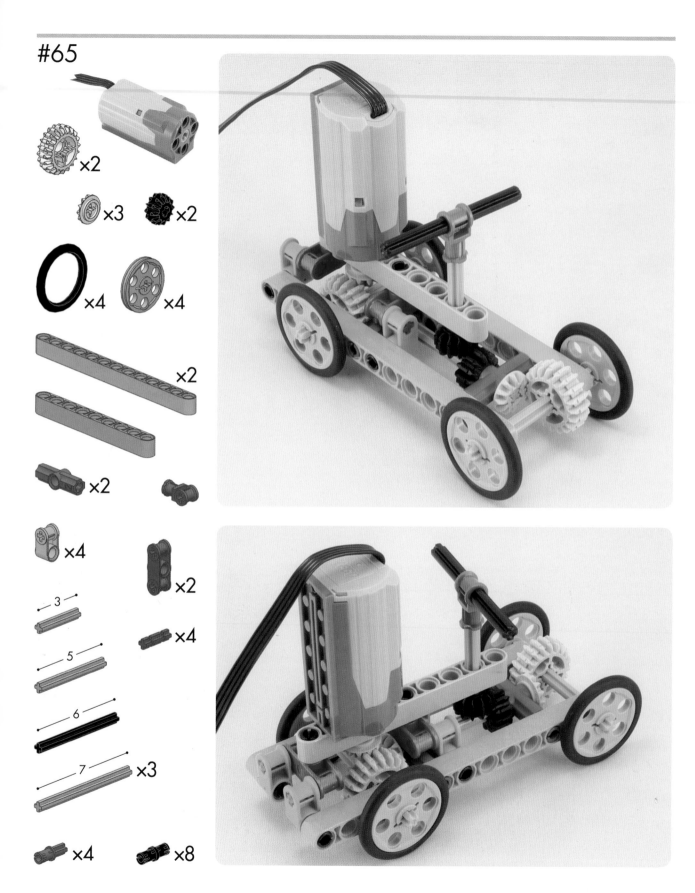

×2

×3 ×2

×4 ×4

×2

×2

×2

×4

×2

3

×4

5

6

7 ×3

×4 ×8

#66

×4

×4

×4

×2 ×2 ×2

×4

×6

×2

×3

×2

×2

×2

4 ×3

5 ×2

8 ×2

12

×2 ×2

×6 ×13

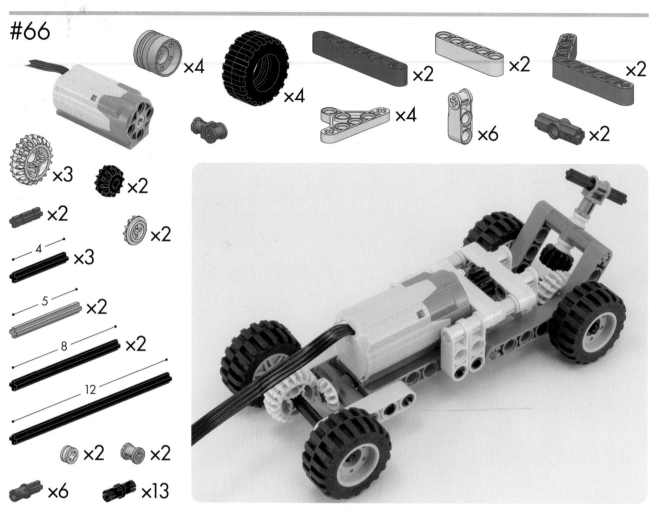

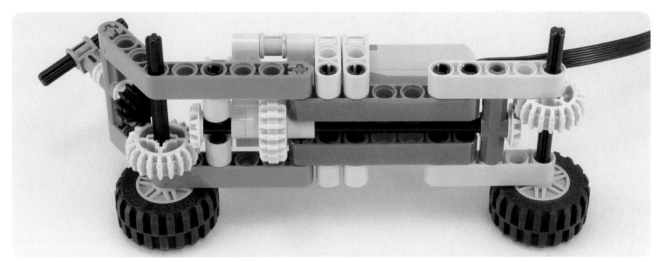

#67

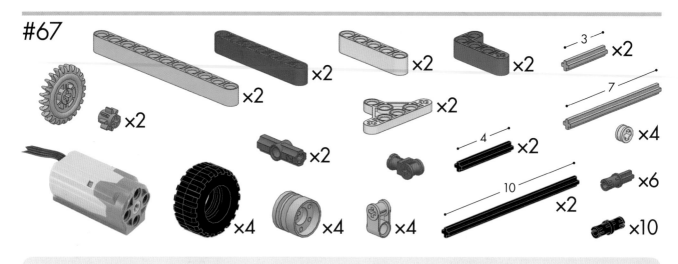

#68

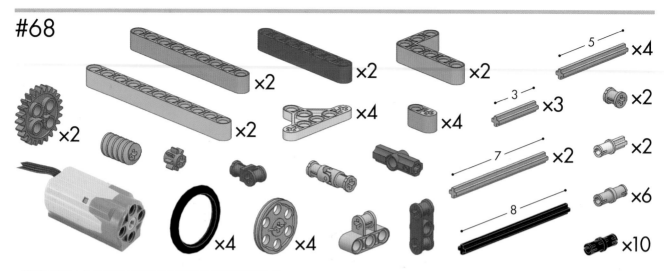

×2 ×2 ×2 ×4 ×2 ×2 ×4 ×4 ×4 ×4 ×2 ×3 ×2 ×4 ×2 ×6 ×10

5 3 7 8

Cars that move something

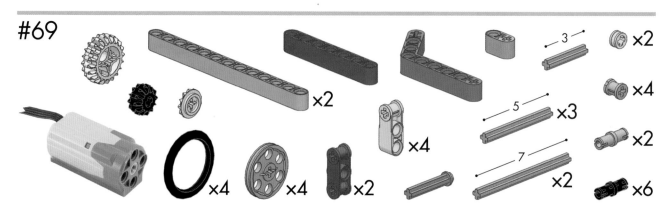

×2 ×4 ×4 ×3 ×2 ×6 ×4 ×4 ×2 ×2

3

5

7

#70

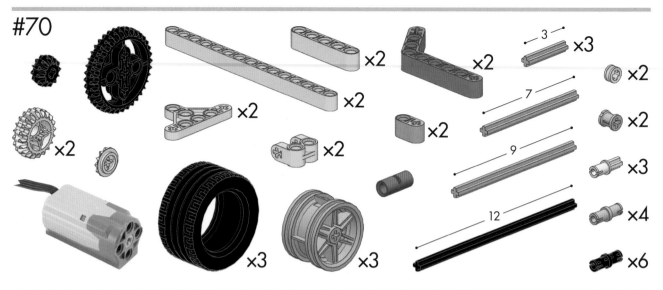

#71

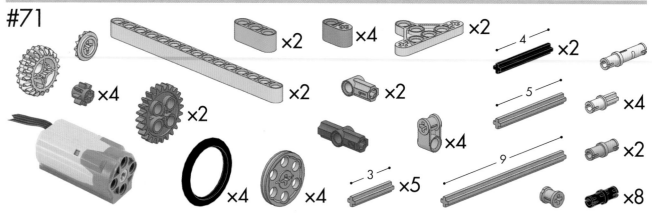

#72

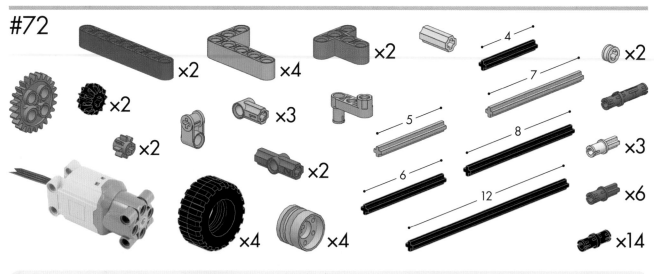

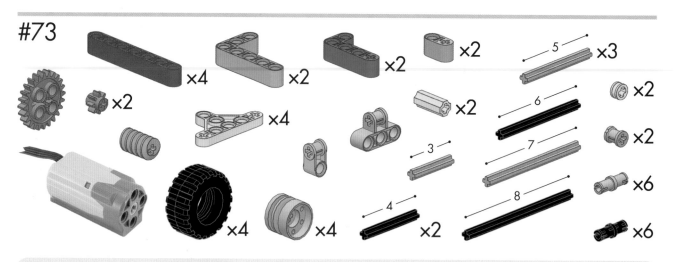

Cars with suspension

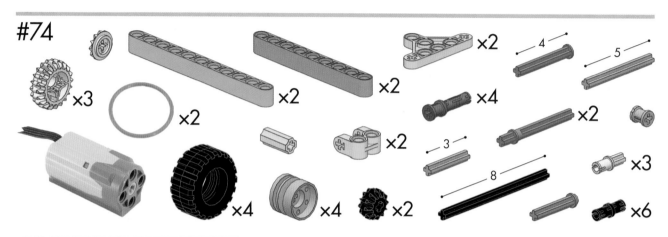

#74

×3
×2
×2
×2
×2
4
×4
5
×2
×2
3
×2
×3
8
×4
×4
×2
×6

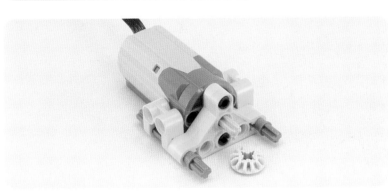

#75

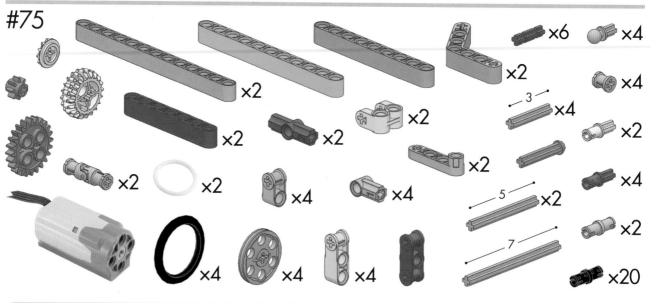

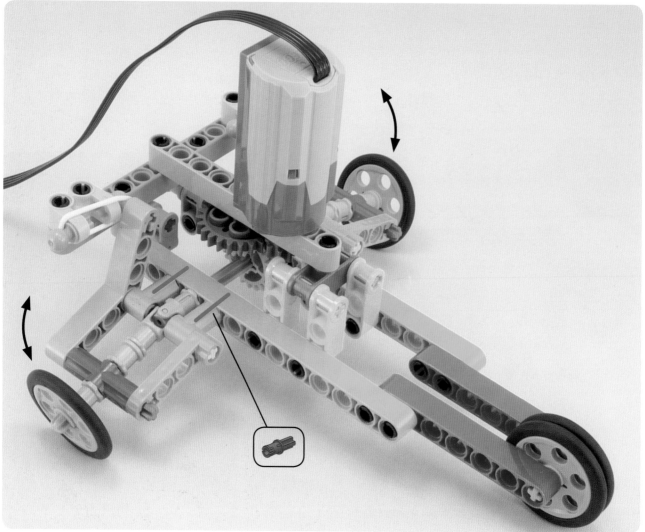

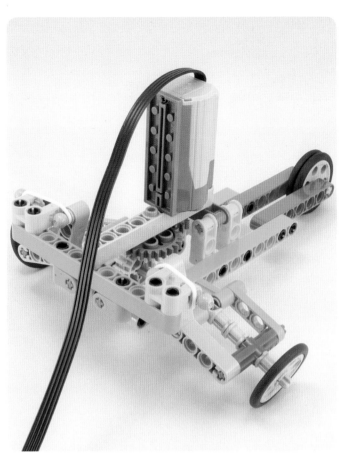
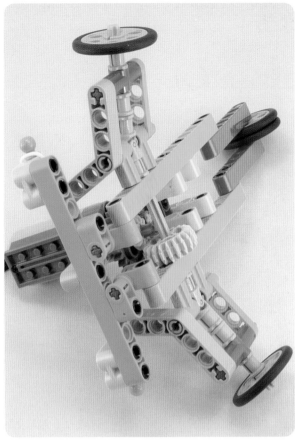

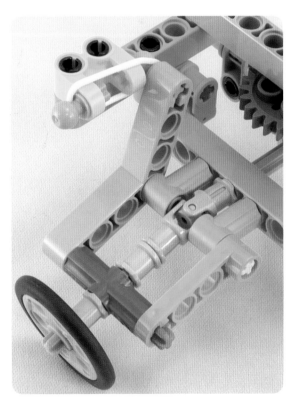

#76

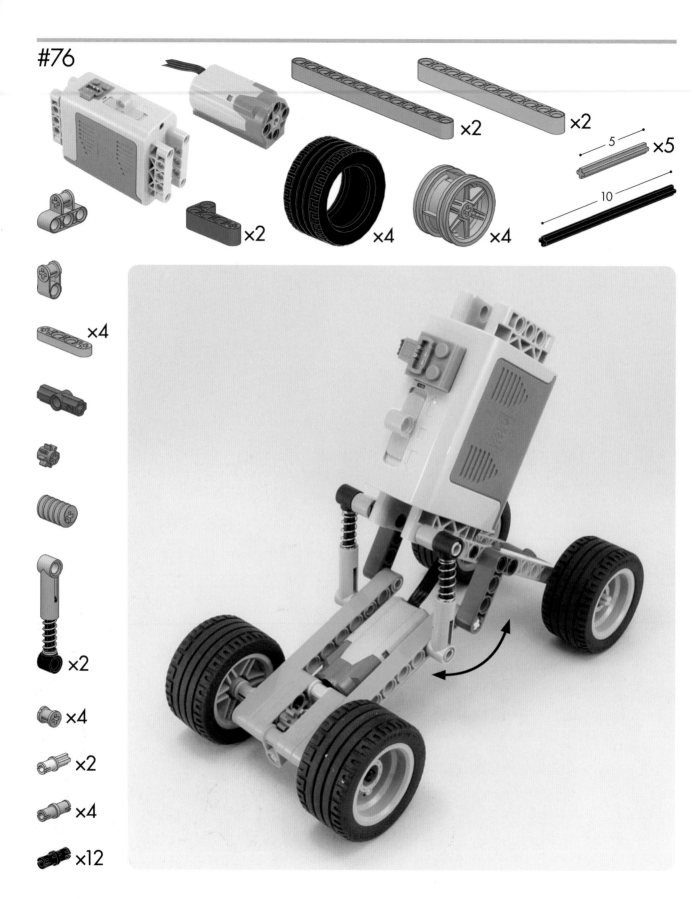

×2 ×2 ×2

5 ×5

10

×2 ×4 ×4

×4

×2

×4

×2

×4

×12

#77

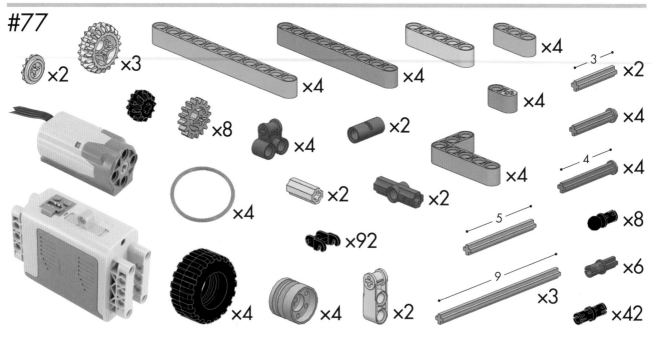

×2　×3　×4　×4　×4　×4　3 ×2　×8　×4　×2　×4　4 ×4　×4　×4　×2　×92　5 ×8　×6　×4　×4　×2　9 ×3　×42

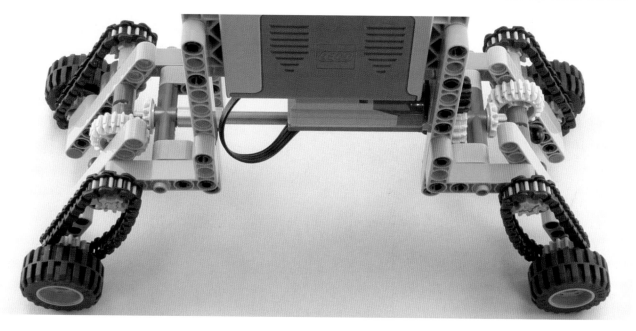

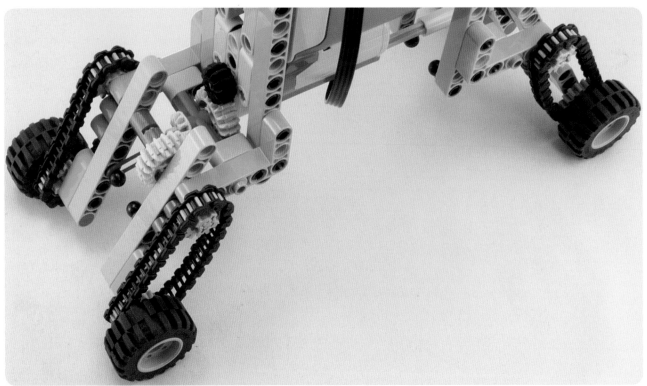

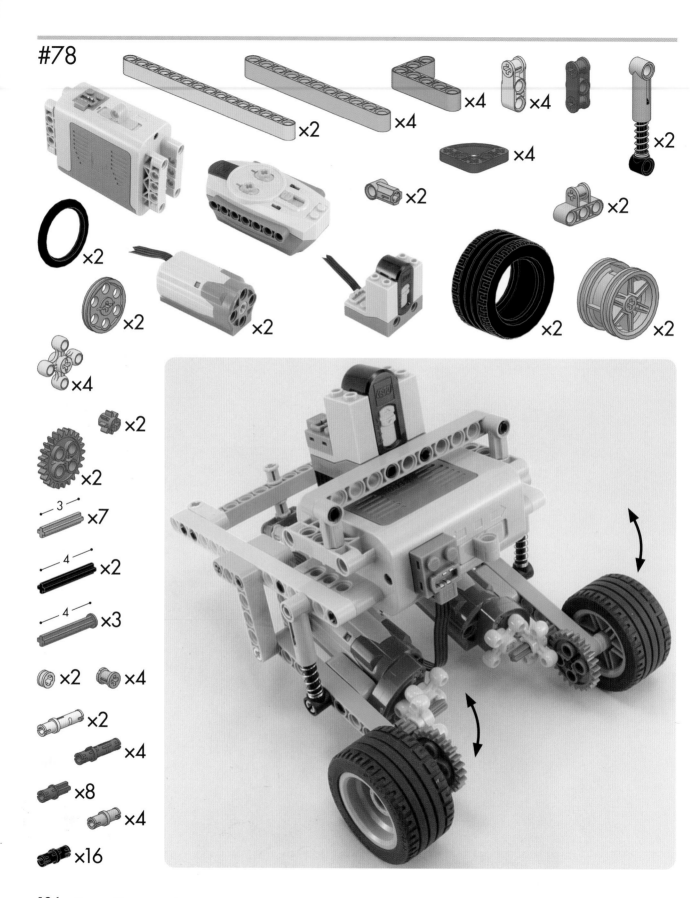

78

×2
×4
×4
×4
×2
×4
×2
×2
×2
×2
×4
×2
×2
×2
×2
×2
×2
×2

×2
3 ×7
4 ×2
4 ×3
×2 ×4
×2
×4
×8
×4
×16

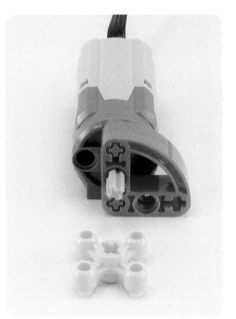

#79

×4
×4
×2
×2
×4
×4
×4
×4
×3
×4
×2
×6
×2
×2
×7
×2
×6
×4
×5 ×2
×14
×4 ×6
×7 ×3
×3 ×3
×4 ×20

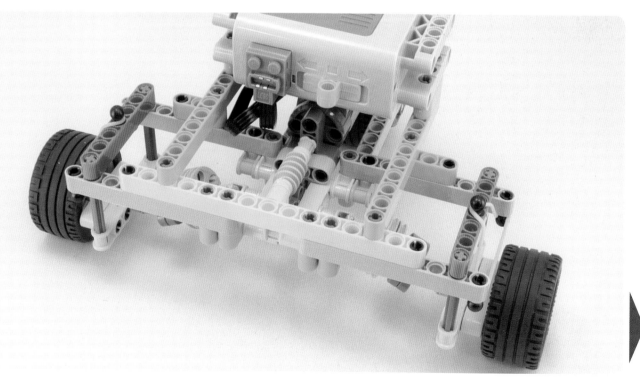

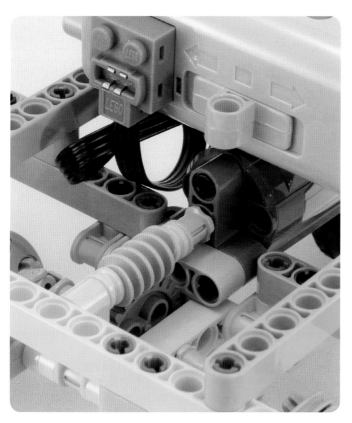

Five different bases for a small car

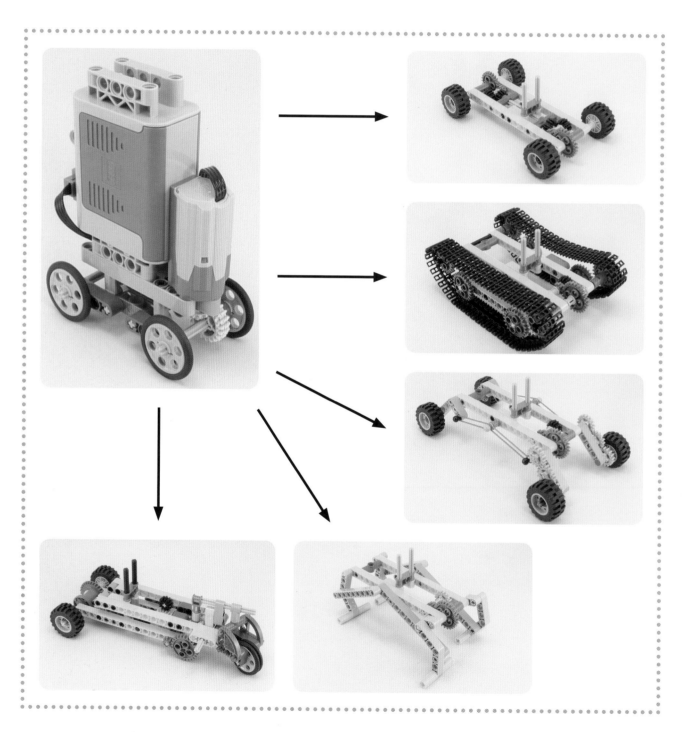

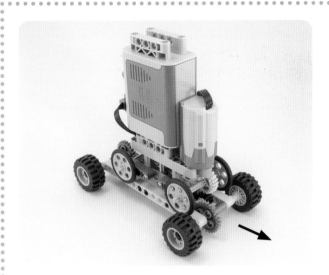

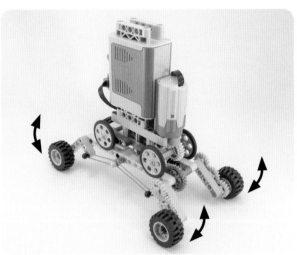

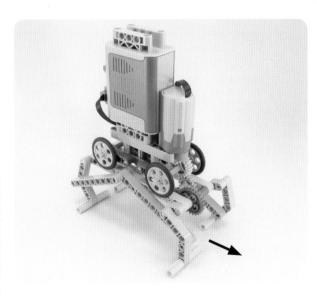

#80

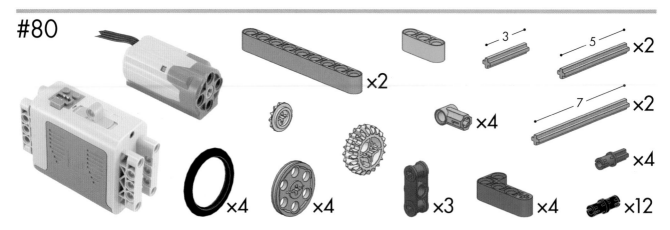

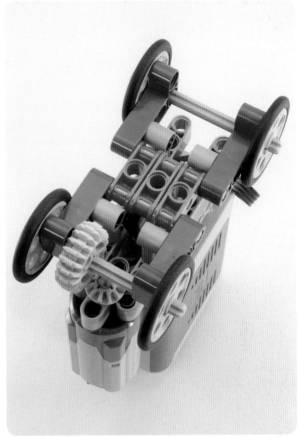

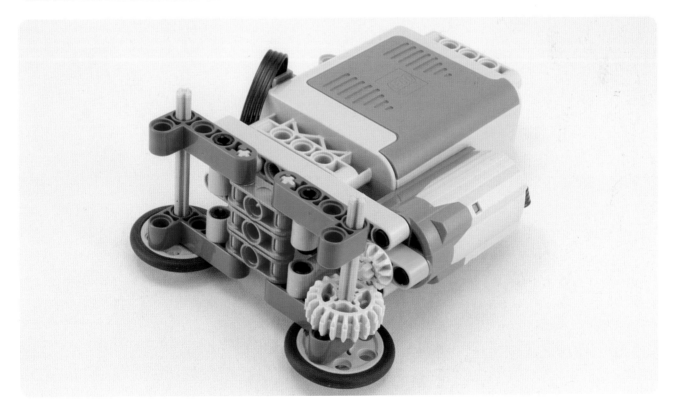

#81

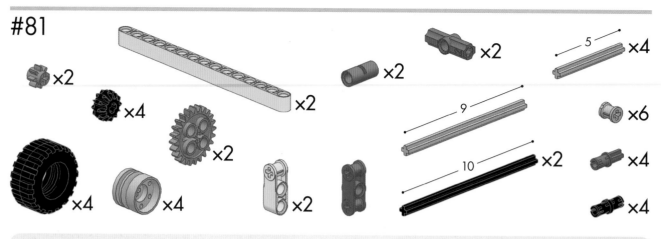

#82

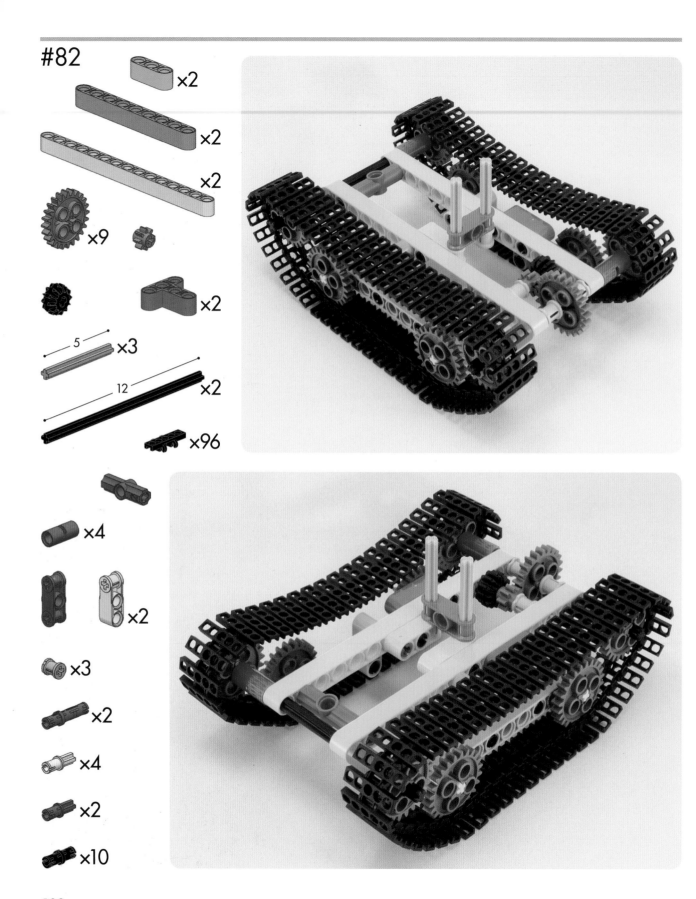

×2
×2
×2
×9
×2
5 ×3
12 ×2
×96
×4
×2
×3
×2
×4
×2
×10

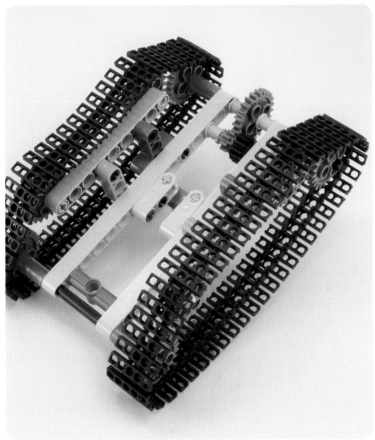

#83

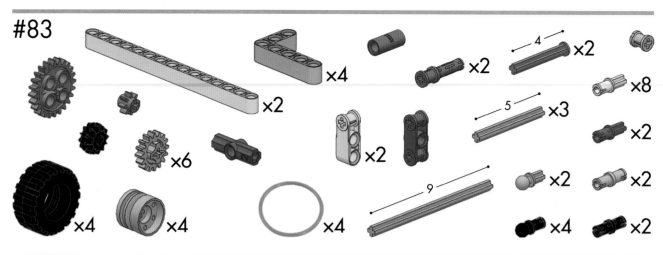

×2 ×4 ×2 ×4 ×2 ×8 ×6 ×2 ×2 ×2 ×4 ×4 ×4 ×3 ×2

#84

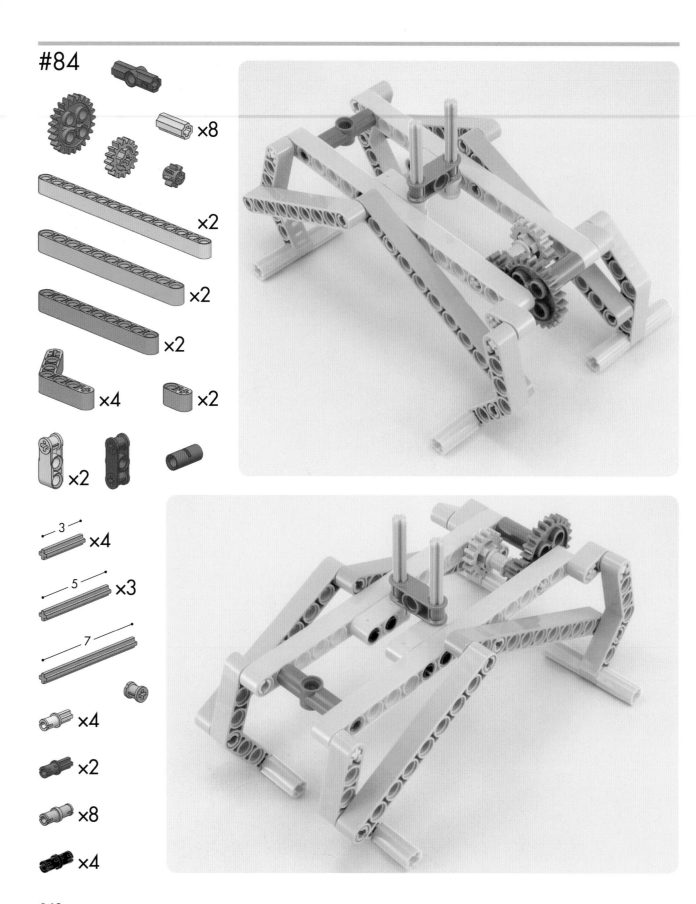

×8

×2

×2

×2

×4 ×2

×2

3 ×4

5 ×3

7

×4

×2

×8

×4

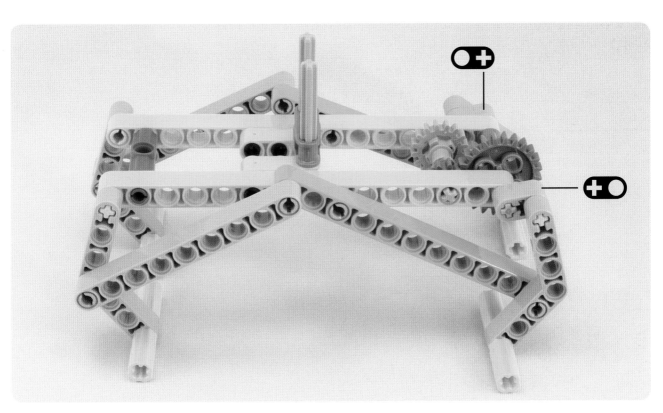

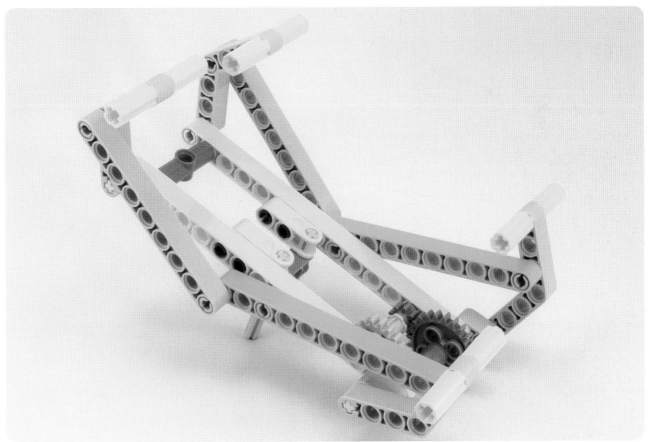

#85

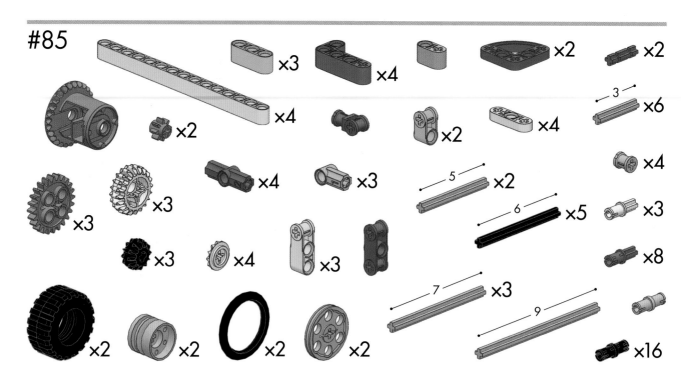

×3 ×4 ×2 ×2

×4 ×4 ×2 ×4 3 ×6

×2 ×2 ×2 ×4

×3 ×3 5 ×2 ×3

×3 ×4 ×3 ×3 6 ×5 ×8

7 ×3

×2 ×2 ×2 ×2 9 ×16

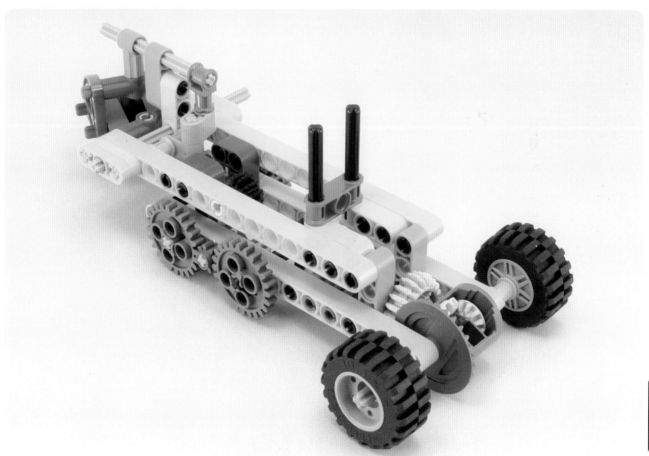

Cars that react

#86

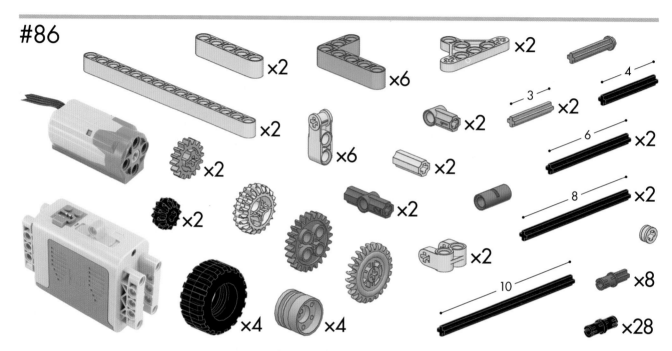

×2
×6
×2
×6
×2
×2
×2
×2
×2
×4
×4
×2
×2
×2
×8
×28

3
4
6
8
10

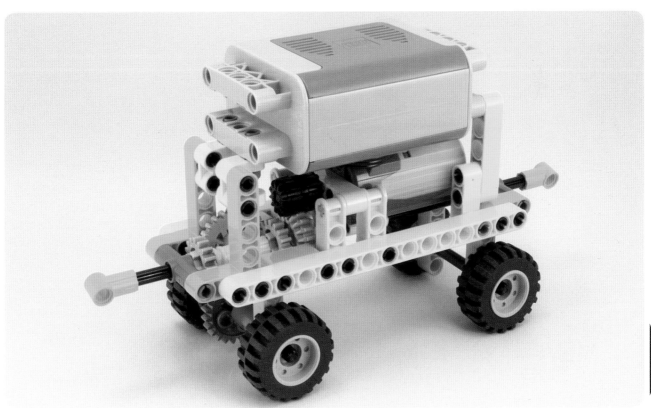

#87

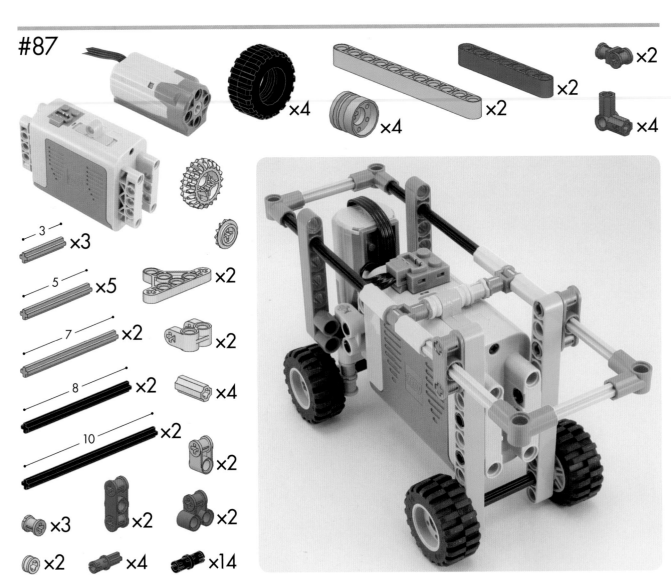

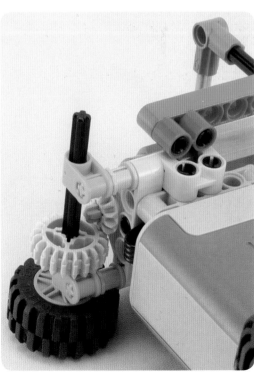

#88

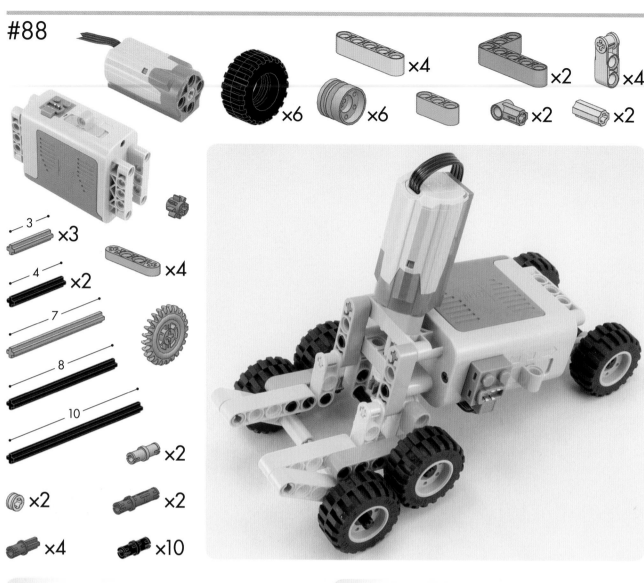

- ×4
- ×2
- ×4
- ×6
- ×6
- ×2
- ×2
- 3 — ×3
- ×4
- 4 — ×2
- 7
- 8
- 10
- ×2
- ×2
- ×2
- ×4
- ×10

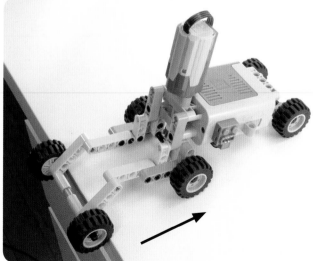

#89

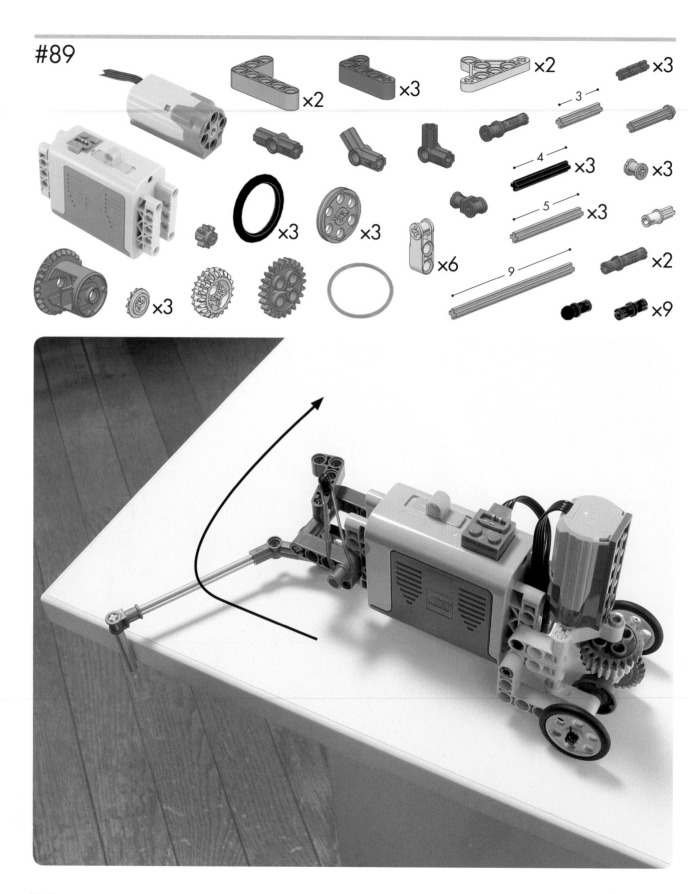

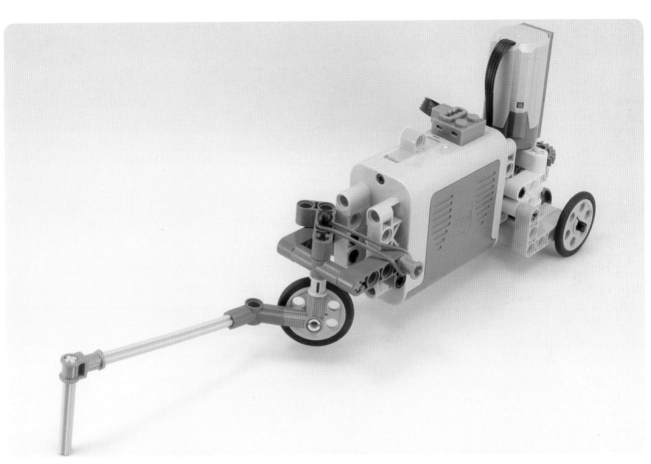

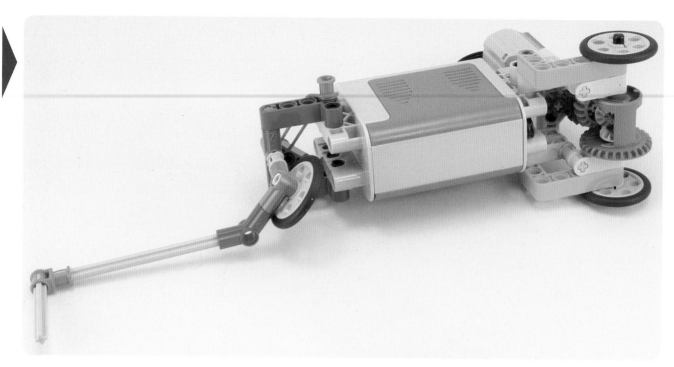

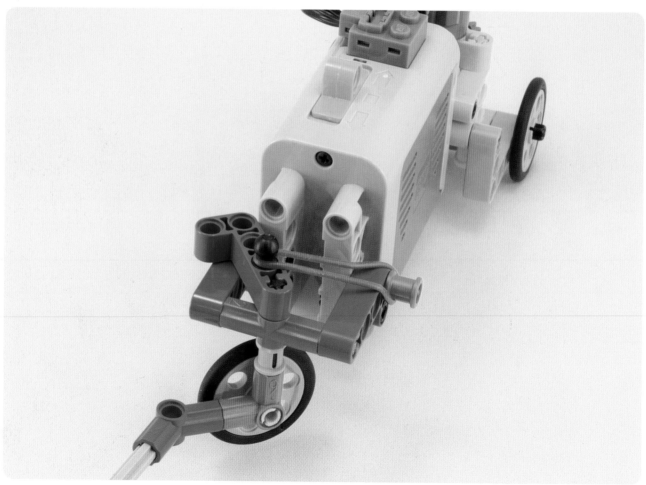

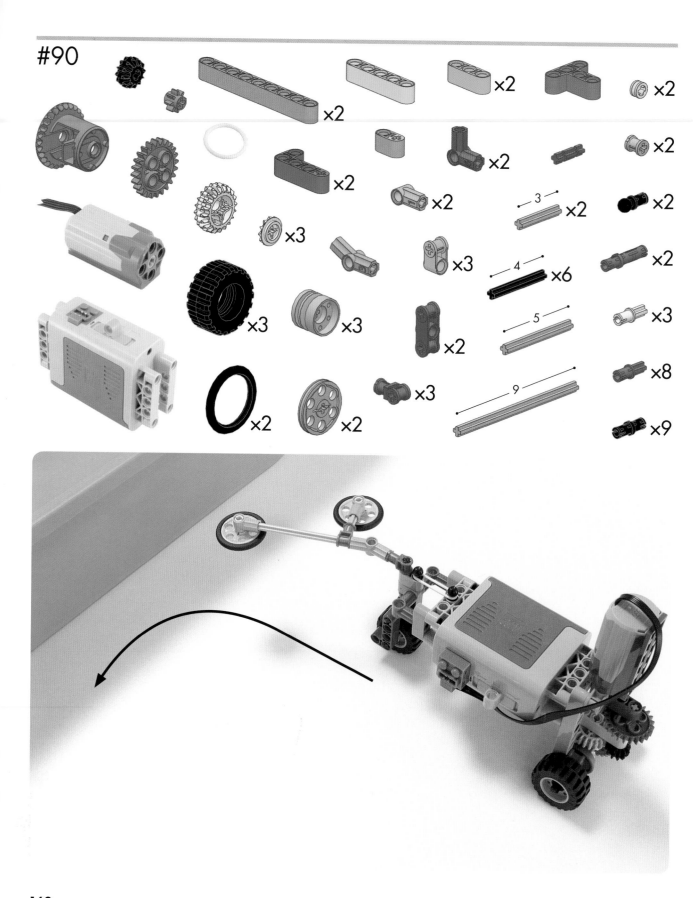

#90

×2 ×2 ×2 ×2

×2

×2

×2

×3

×3 ×3 ×6 ×2

×3 ×2 ×2

×3 ×8

×2 ×2 ×3 ×9

3

4

5

9

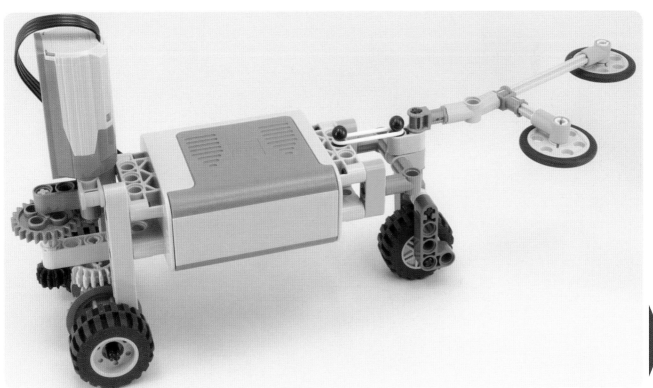

#91

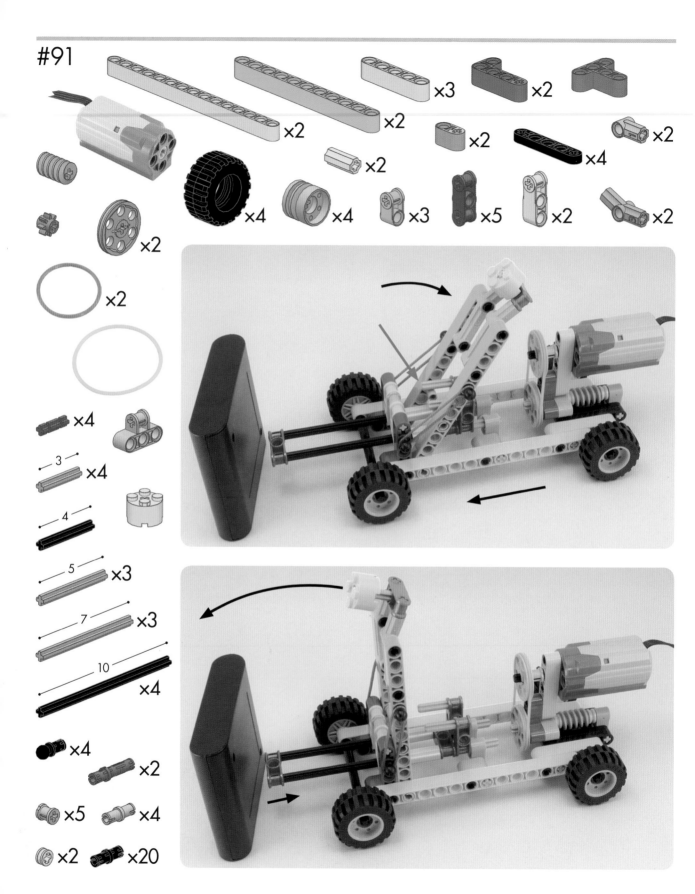

#92

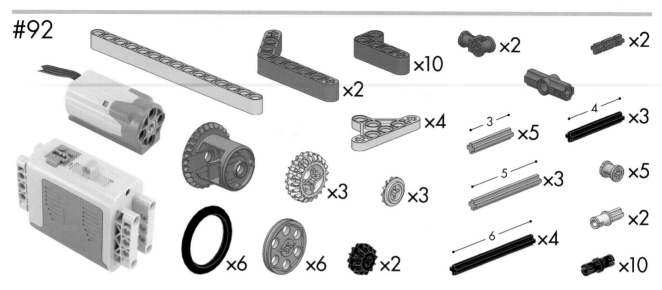

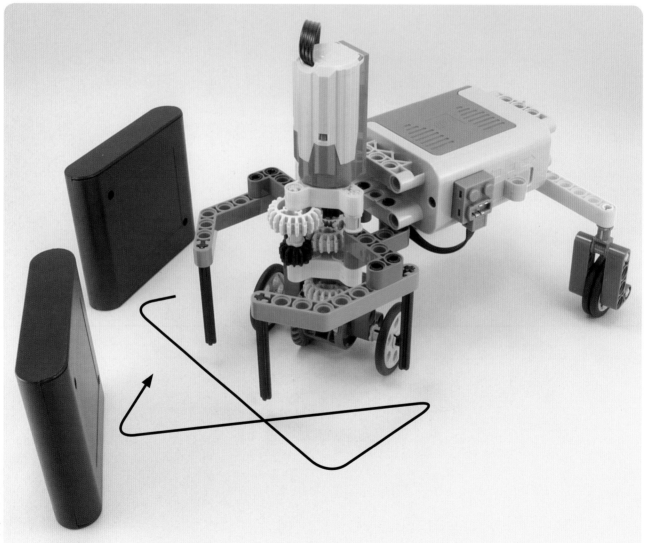

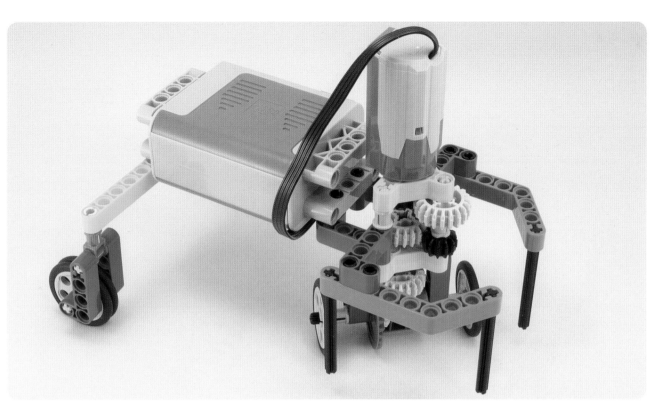

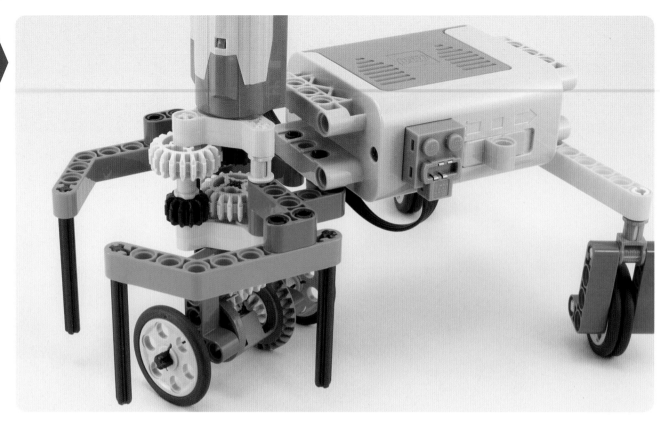

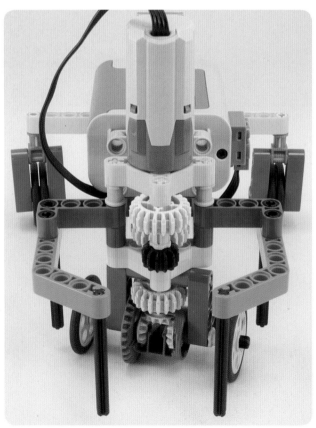

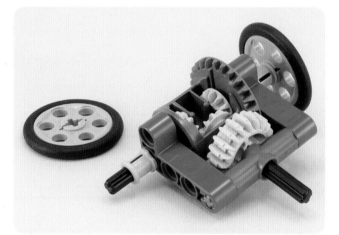

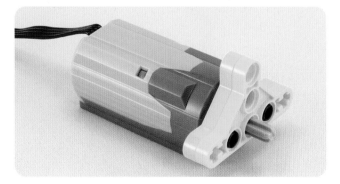

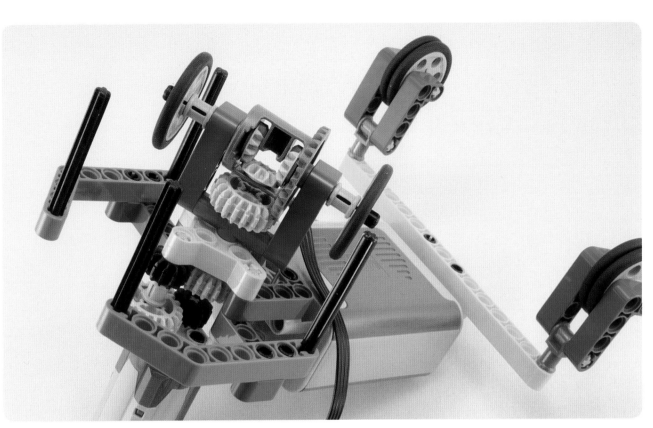

Cool cars

#93

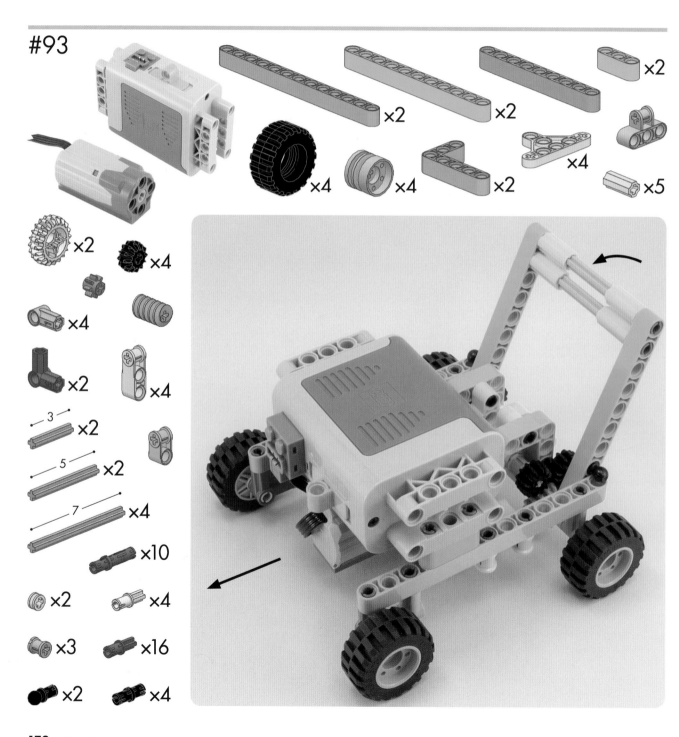

×2 ×2 ×2 ×2

×4 ×4 ×2 ×4 ×5

×2 ×4

×4

×2 ×4

3 ×2

5 ×2

7 ×4

×10

×2 ×4

×3 ×16

×2 ×4

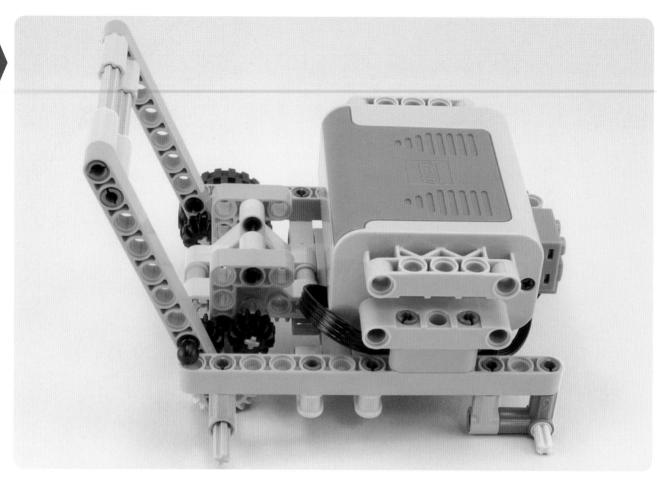

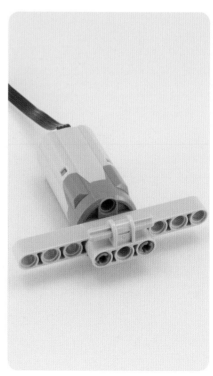

#94

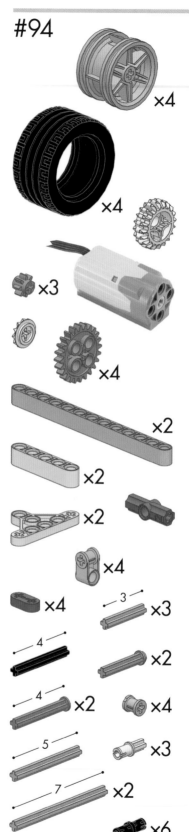

×4

×4

×3

×4

×2

×2

×2

×4

×4

3 ×3

4

4 ×2

5

7 ×2

×2

×4

×3

×6

#95

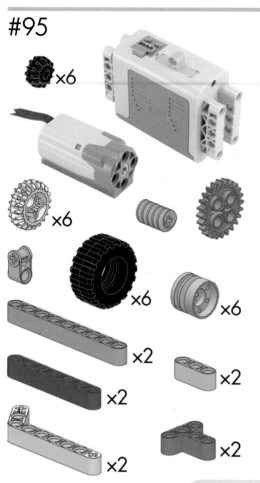

×6

×6

×6

×6

×6

×2

×2

×2

×2

×4

×2

×6

×6

×6

×2

3

5

6

9

×4

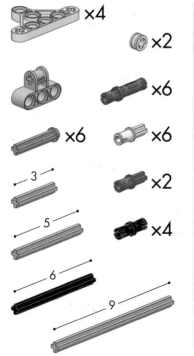

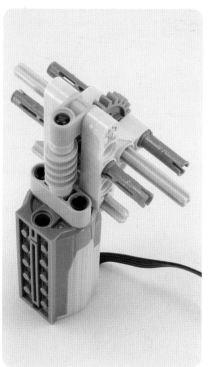

#96

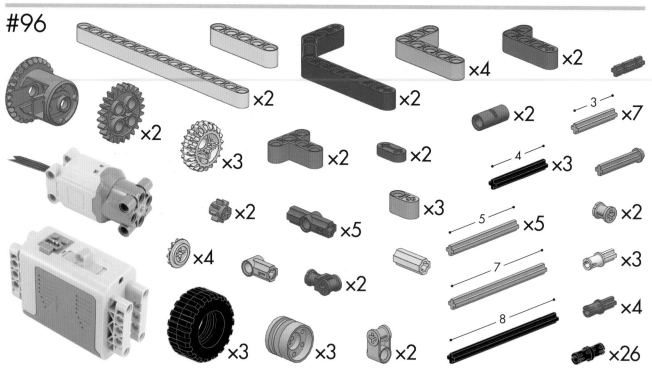

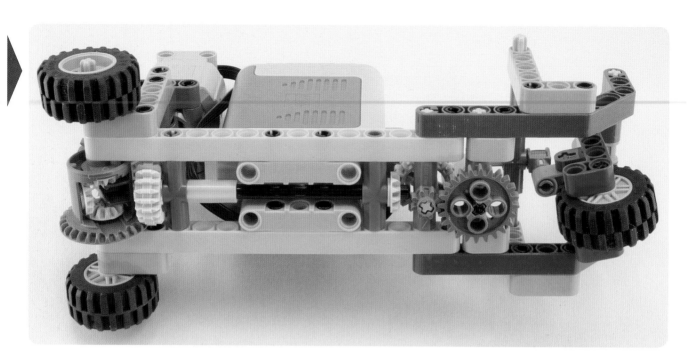

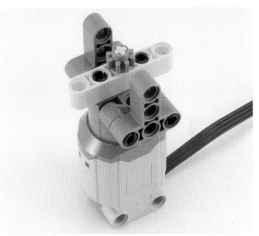

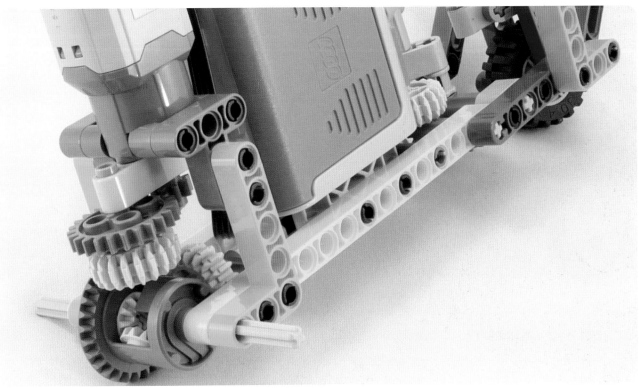

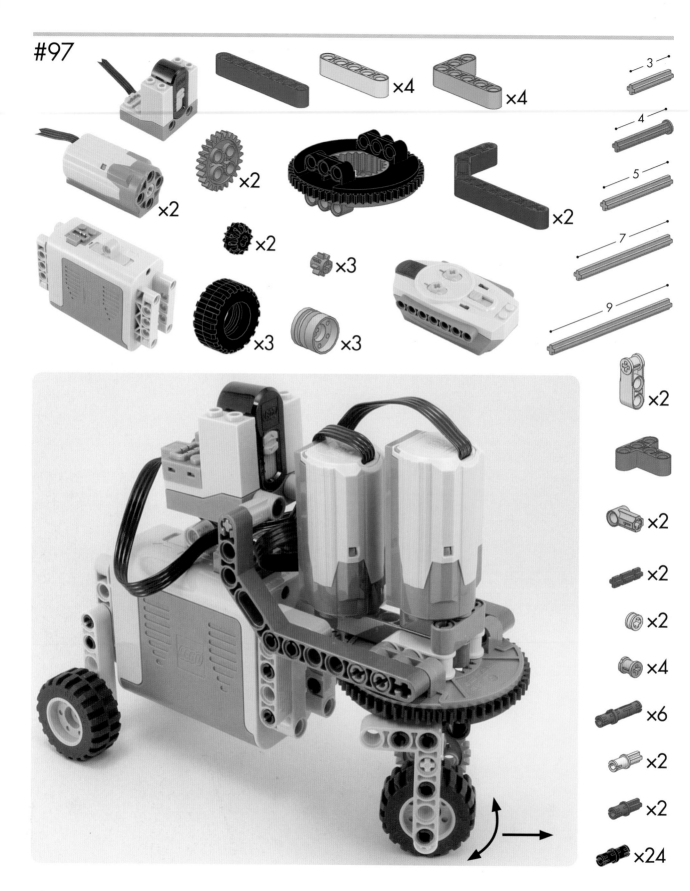

×2 ×4 ×4

3

×2 4

×2 5

×2 ×2 7

×2 ×3 9

×2 ×3 ×3

×2

×2

×2

×2

×4

×6

×2

×2

×24

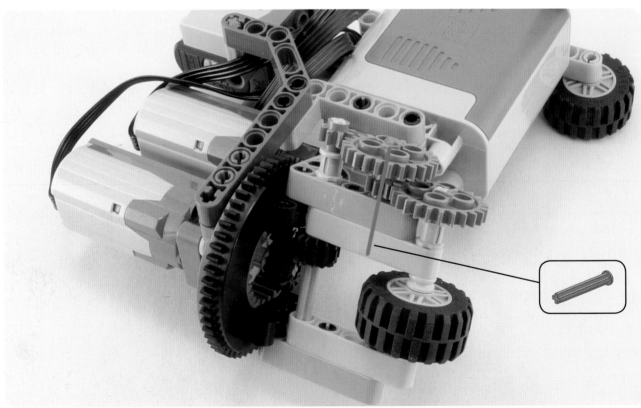

#98

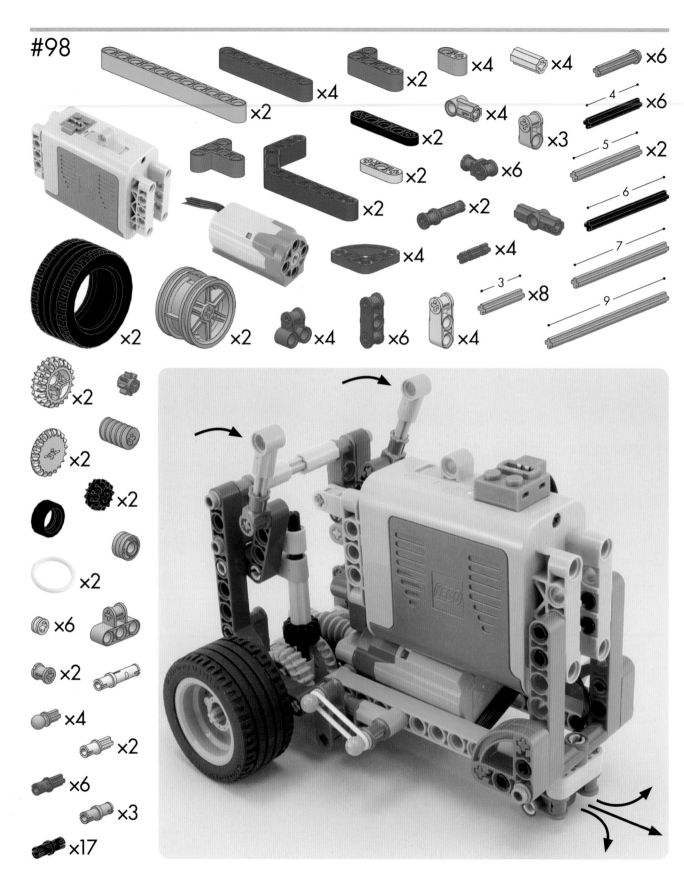

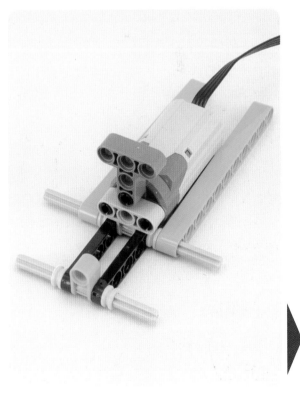

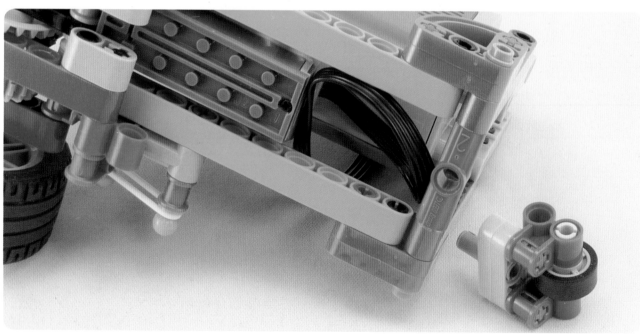

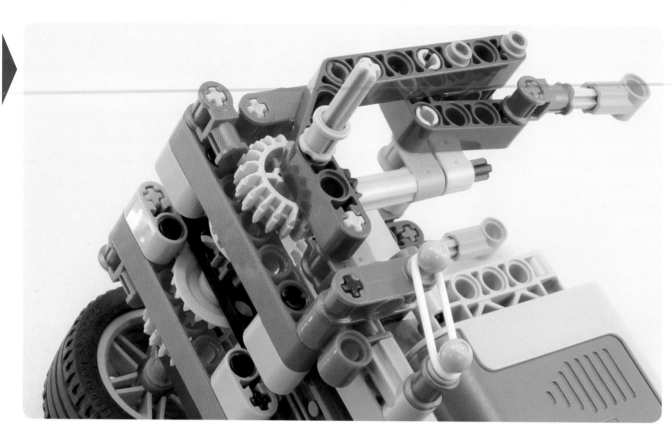

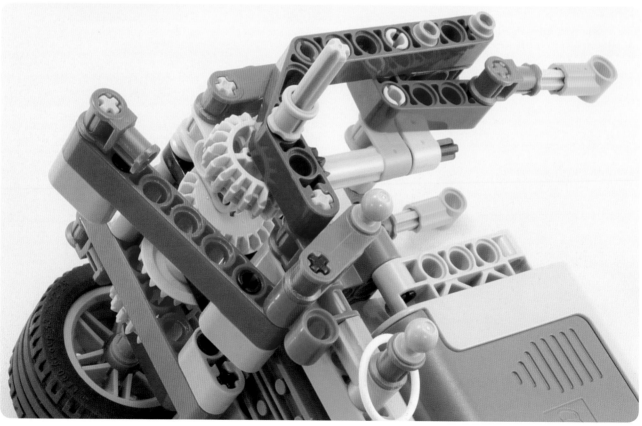

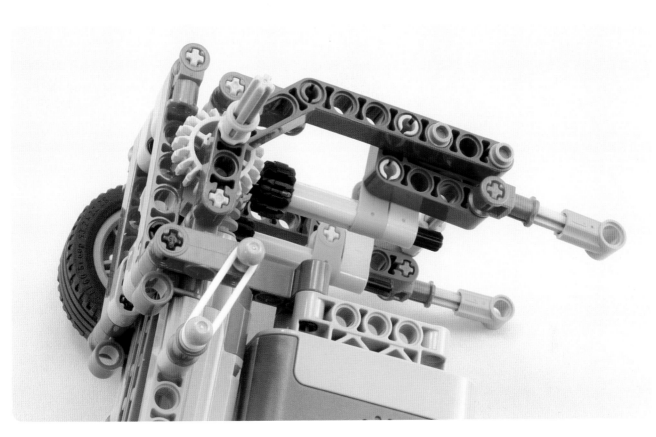

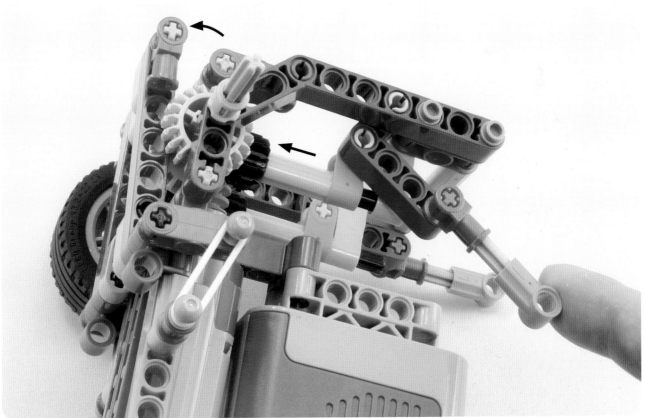

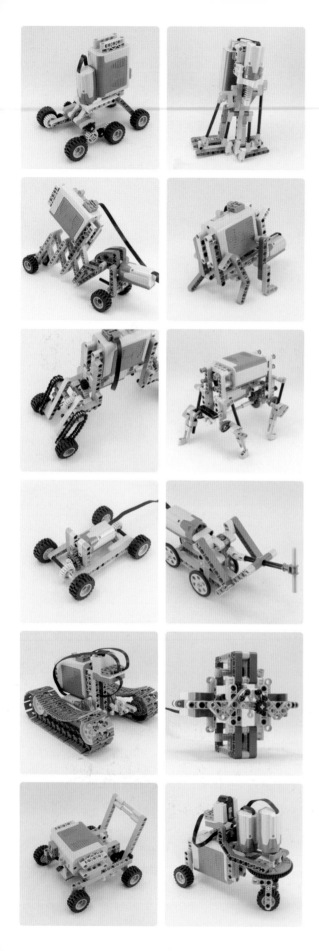

PART 2

Moving Without Tires

196

220

234

244

208

224

240

Two-legged walkers

99

×2

×6

×6

×2

×4

×4

3 ×2

6 ×5

7

×2

×4

×4

×10

×10

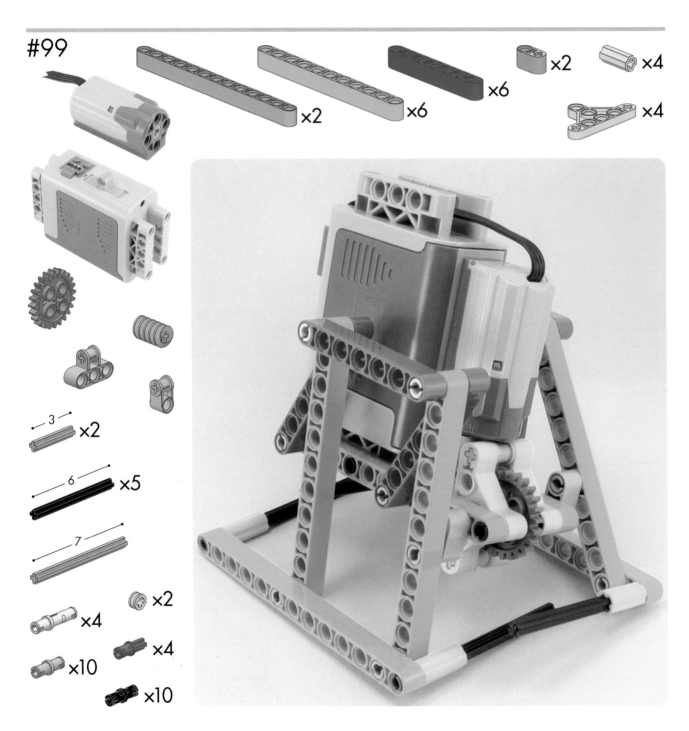

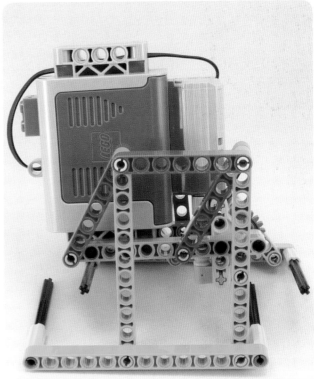

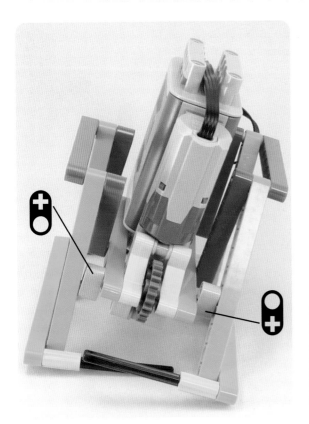

#100

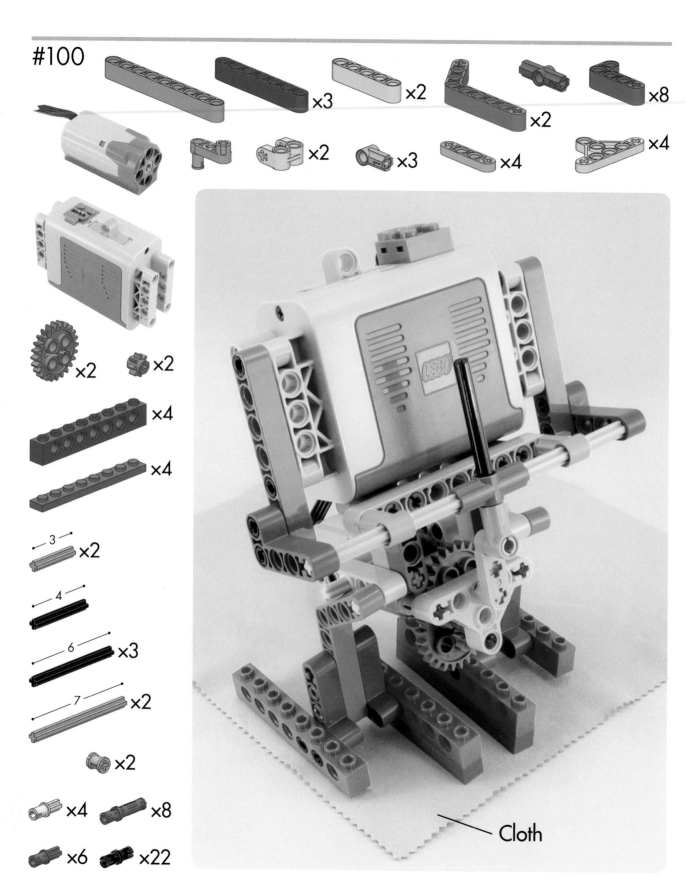

×3 ×2 ×2 ×8

×2 ×2 ×3 ×4 ×4

×2 ×2

×4

×4

3 ×2

4

6 ×3

7 ×2

×2

×4 ×8

×6 ×22

Cloth

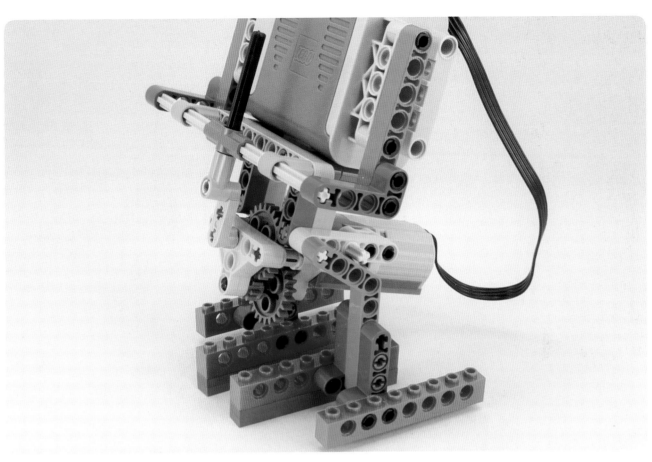

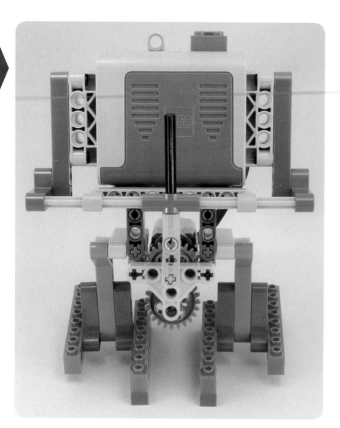

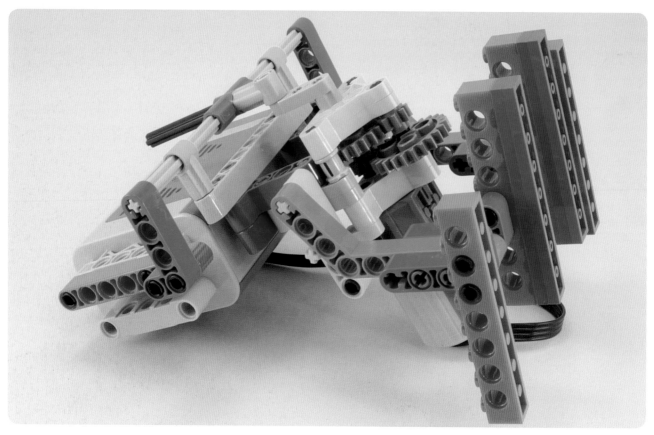

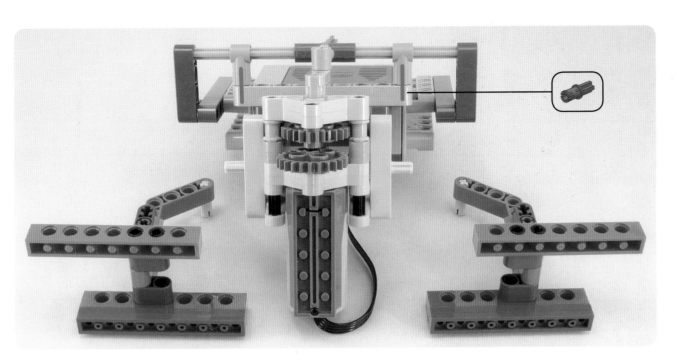

#101

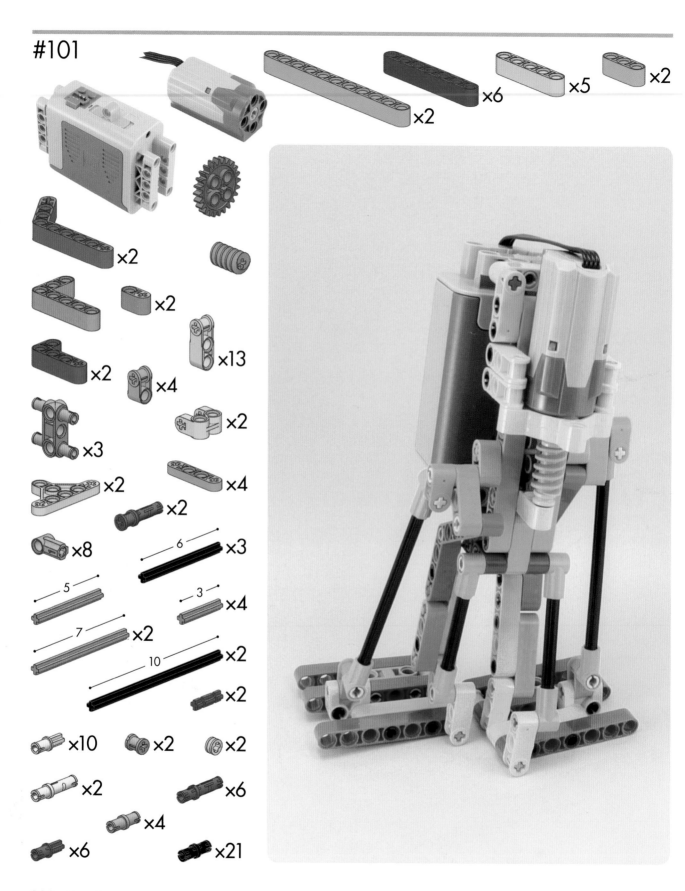

×2 ×6 ×5 ×2

×2 ×2 ×13 ×4 ×2 ×3 ×2 ×4 ×2 ×8

6 ×3
5
3 ×4
7 ×2
10 ×2
×2

×10 ×2 ×2 ×2 ×6 ×4 ×6 ×21

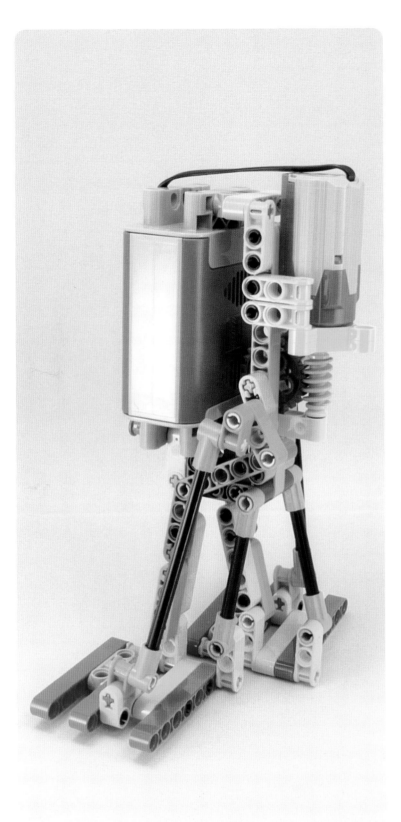
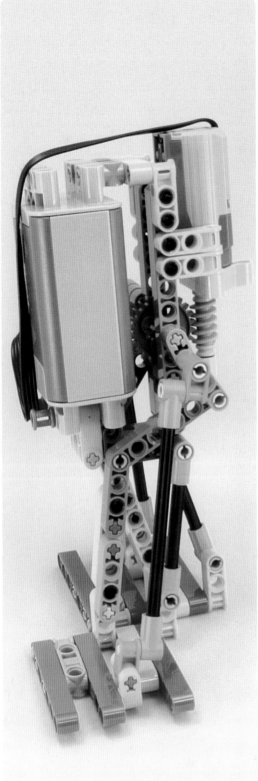

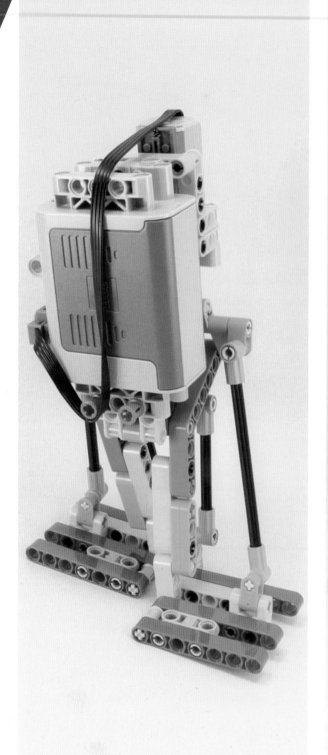

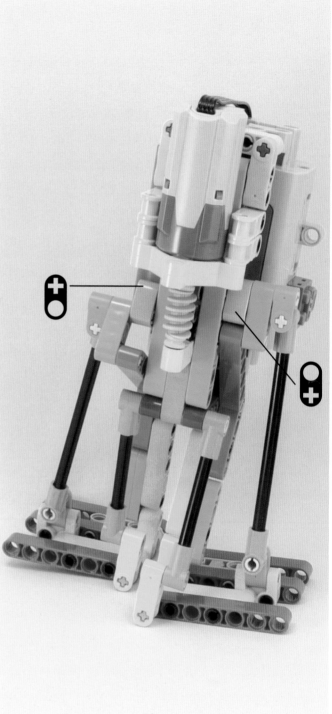

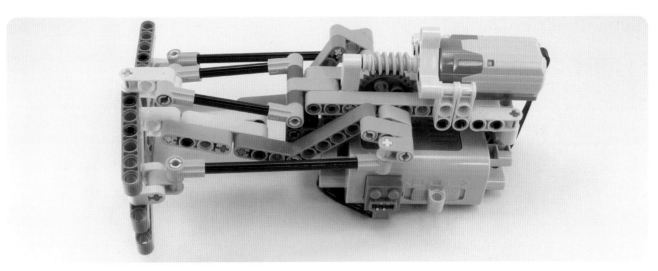

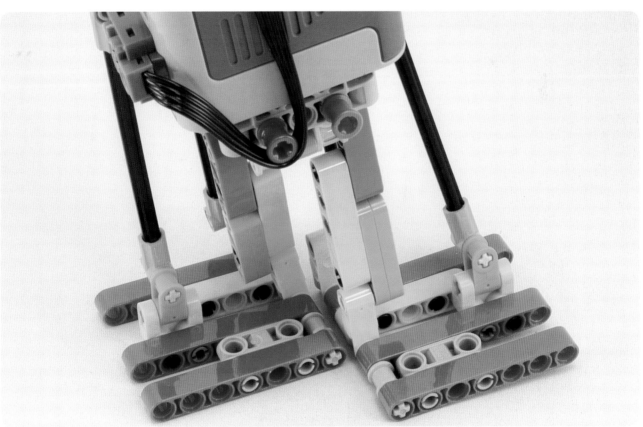

Four-legged walkers

#102

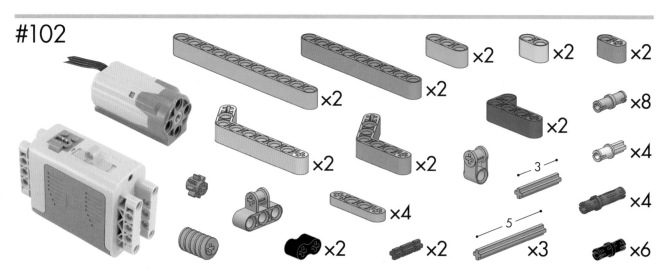

×2 ×2 ×2 ×2 ×2

×2 ×2 ×2 ×8

×4 ×4

3

×2 ×4

5

×2 ×2 ×3 ×6

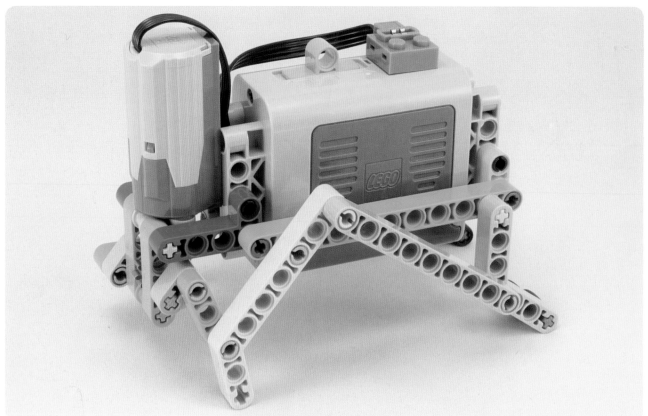

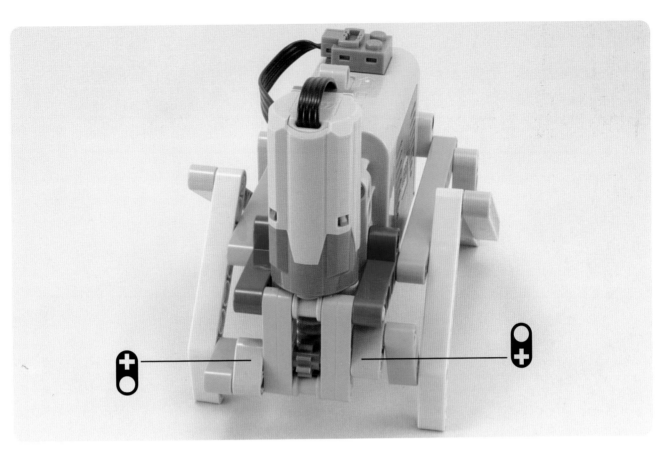

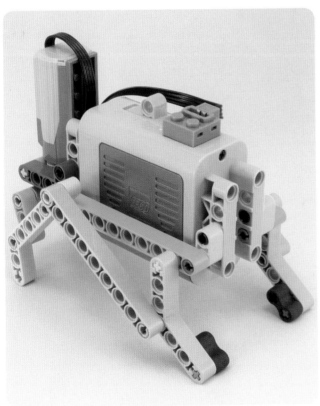

#103

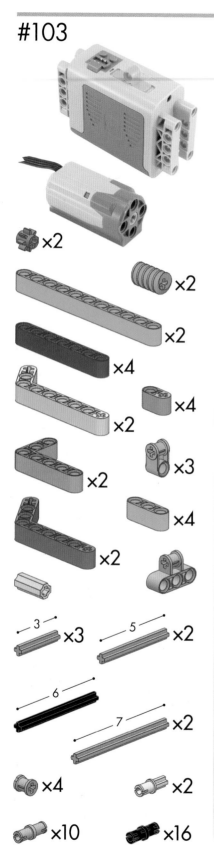

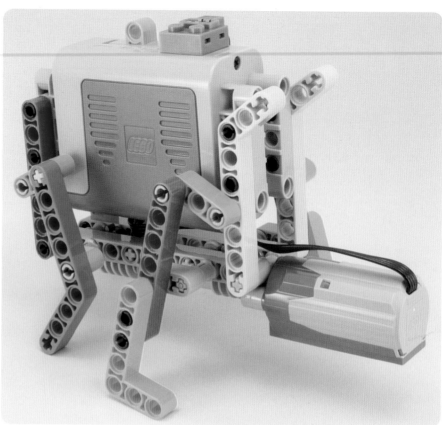

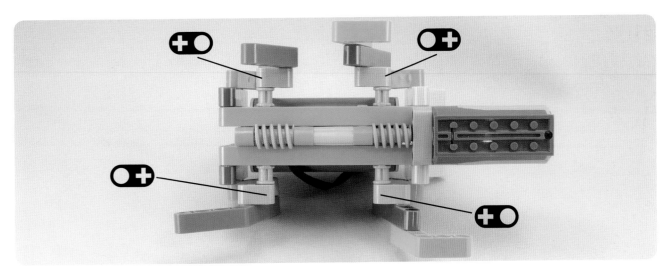

#104

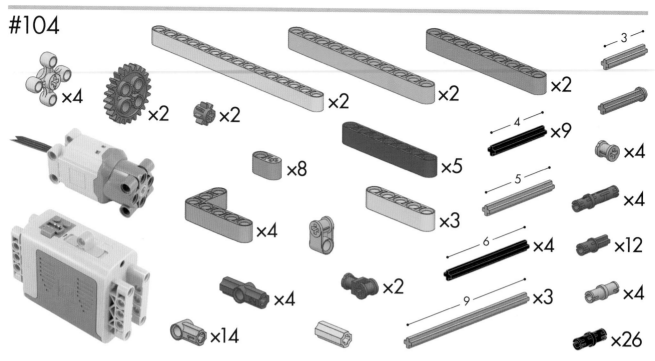

×4 ×2 ×2 ×2 ×2 ×2 ×3

×8 ×9 ×4

×5 ×4

×3 ×4 ×4

×14 ×4 ×2 ×3 ×12 ×4 ×26

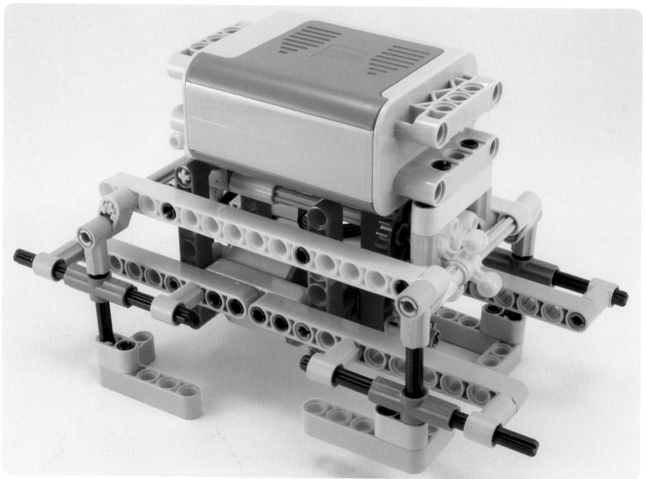

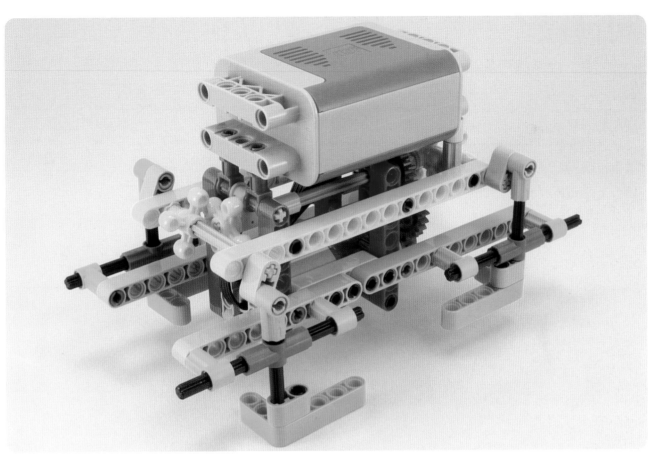

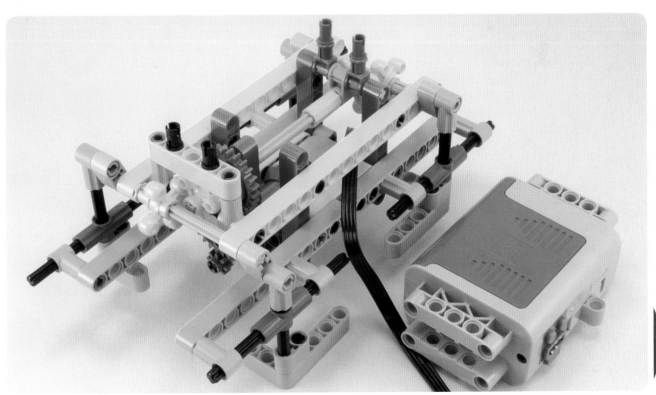

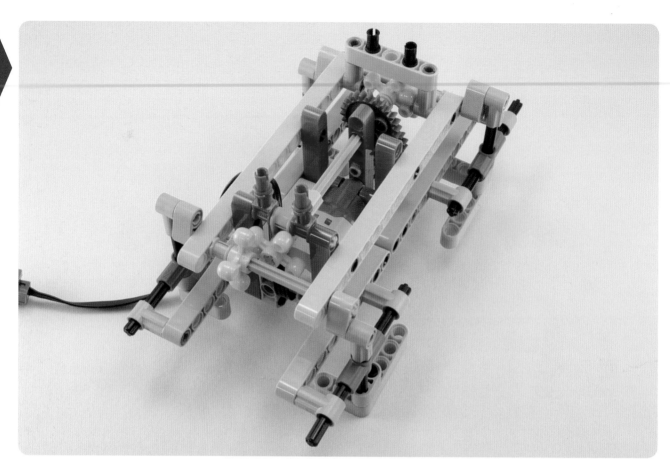

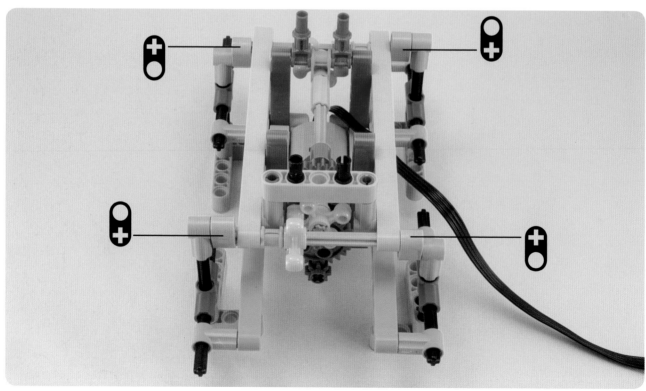

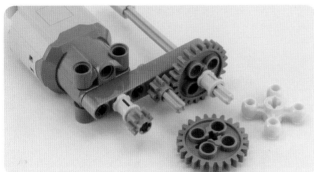

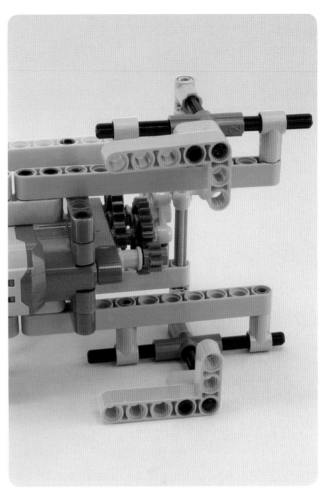

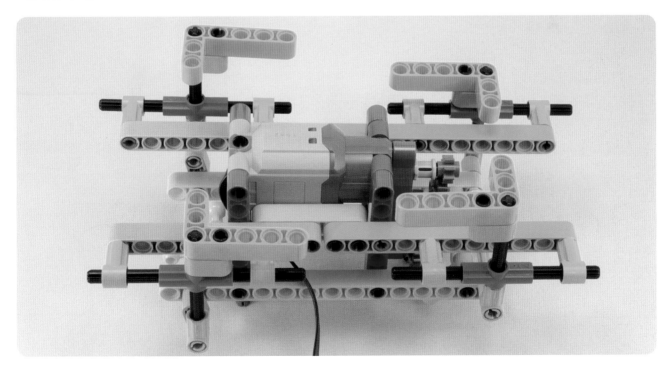

#105

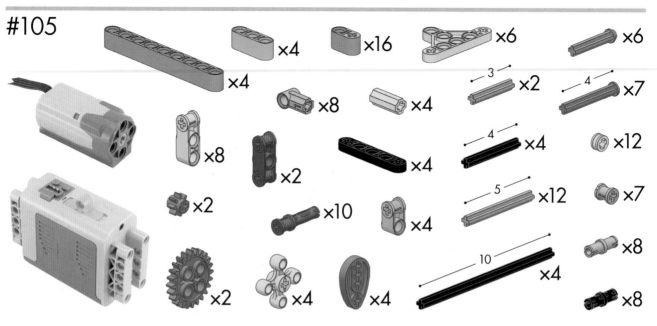

×4 ×4 ×16 ×6 ×6

×8 ×8 ×4 3 ×2 4 ×7

×2 ×4 4 ×4 ×12

×2 ×10 ×4 5 ×12 ×7

×2 ×4 ×4 10 ×4 ×8

×8

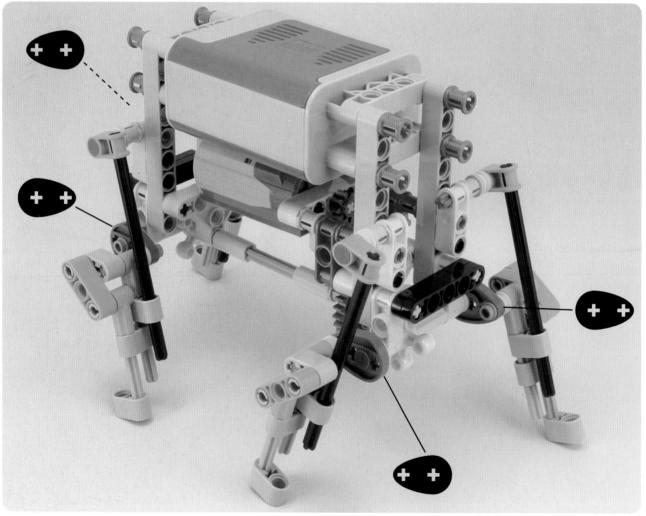

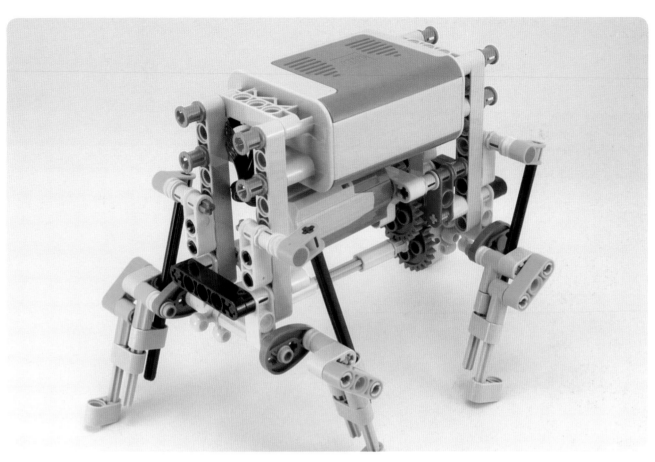

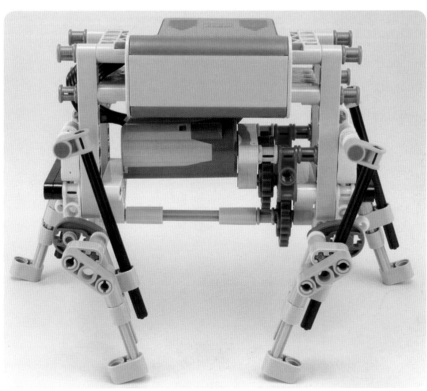

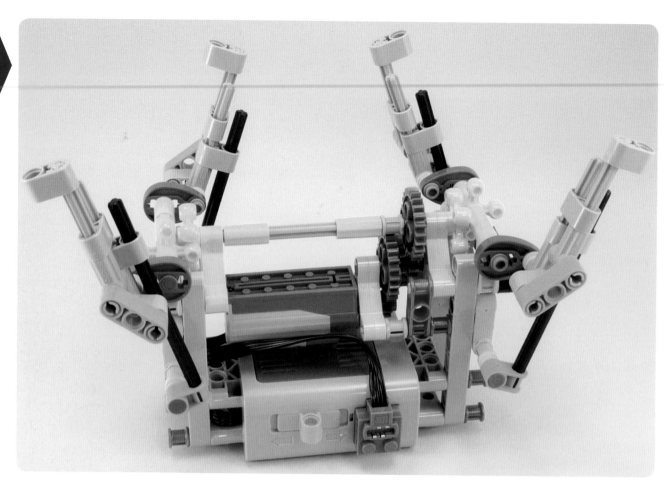

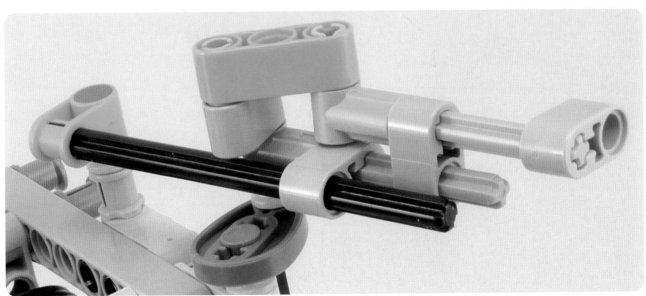

Six-legged walkers

#106

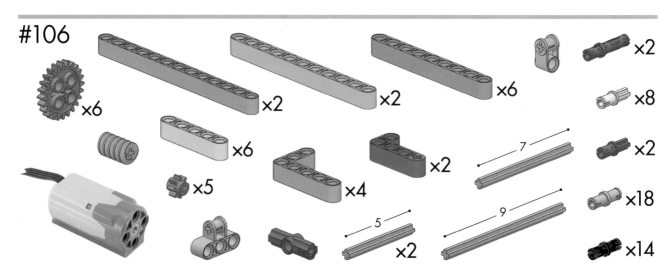

×6 ×2 ×2 ×6 ×2

×6 ×8

×5 ×4 ×2 7 ×2

5 9 ×18

×2 ×14

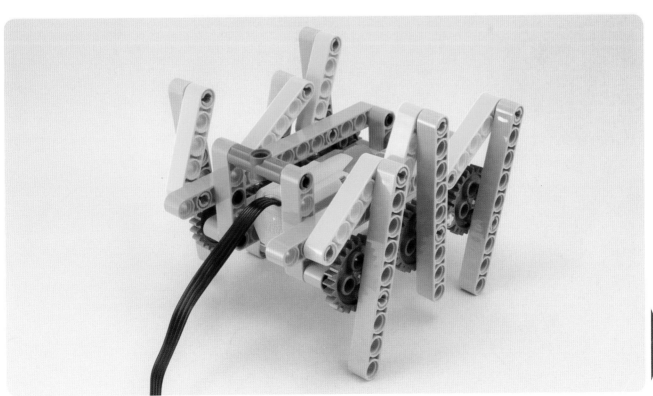

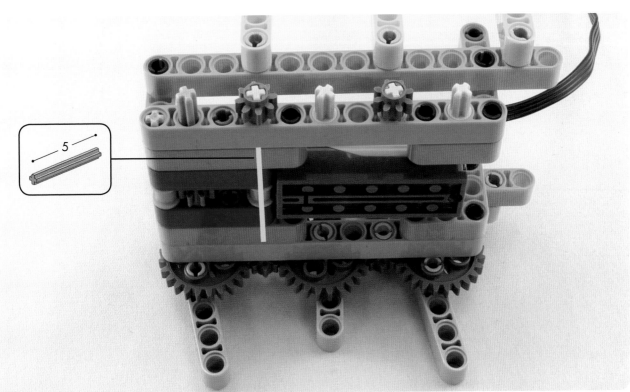

Cool walkers

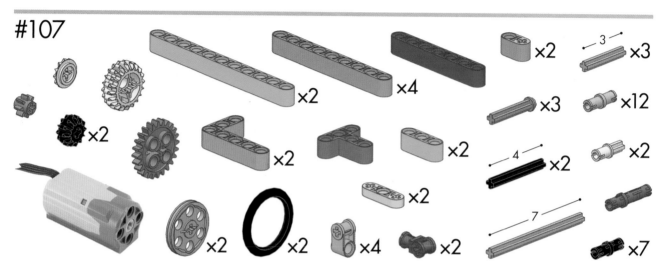

#107

×2 ×4 ×2 — 3 — ×3

×3 ×12

×2 ×2 — 4 — ×2 ×2

×2 ×2 ×4 ×2 — 7 — ×7

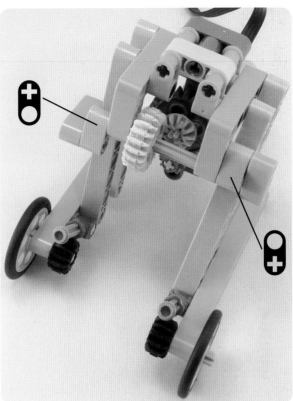

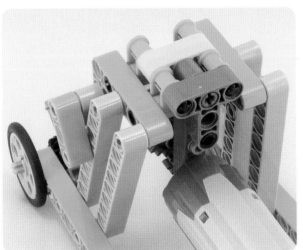

#108

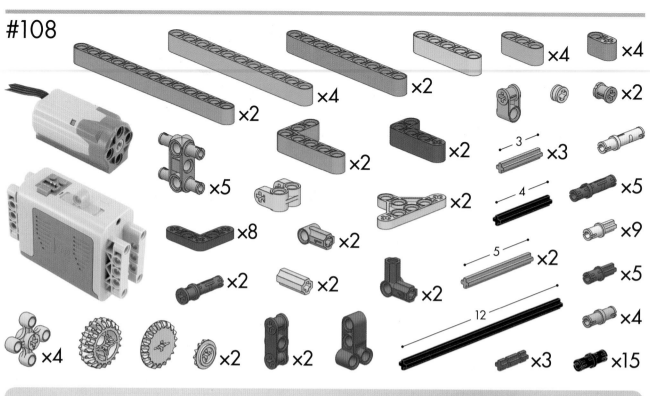

×2 ×4 ×2 ×4 ×4 ×2 ×5 ×2 ×3 ×3 ×5 ×2 ×8 ×2 ×2 ×9 ×2 ×2 ×2 ×5 ×4 ×2 ×2 ×3 ×4 ×15 ×12

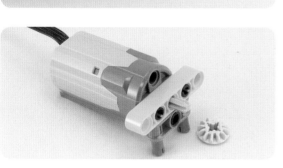

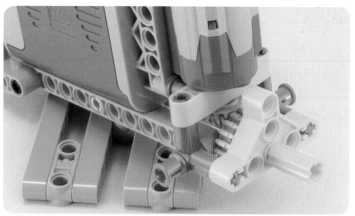

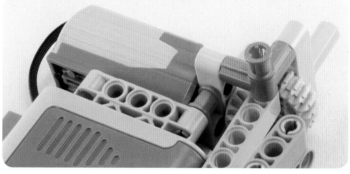

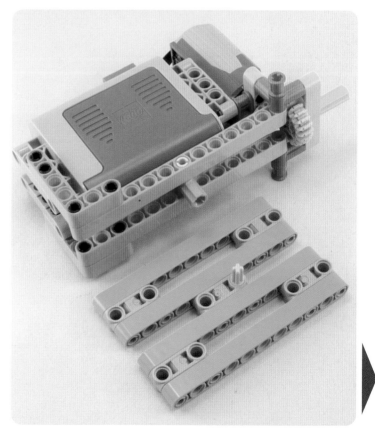

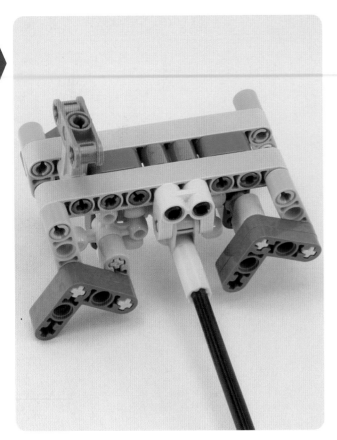

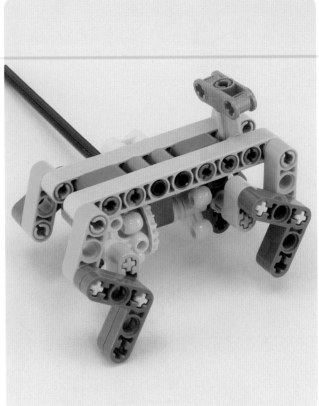

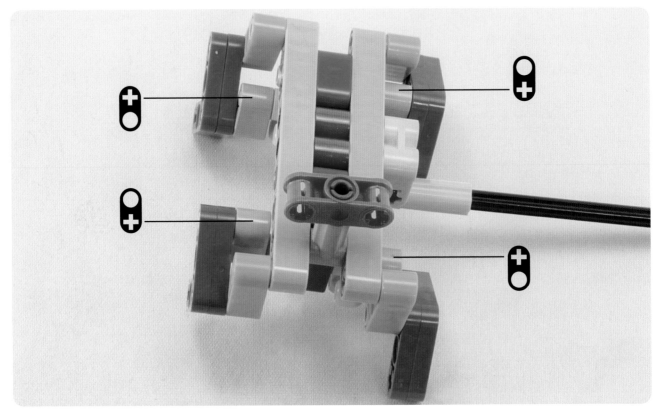

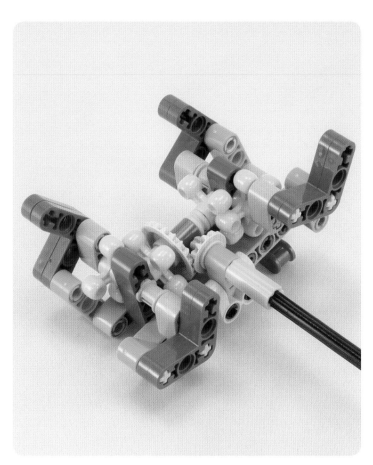

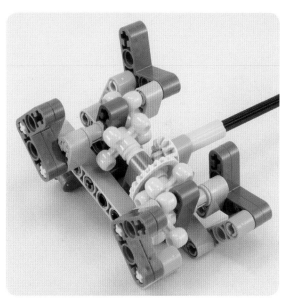

#109

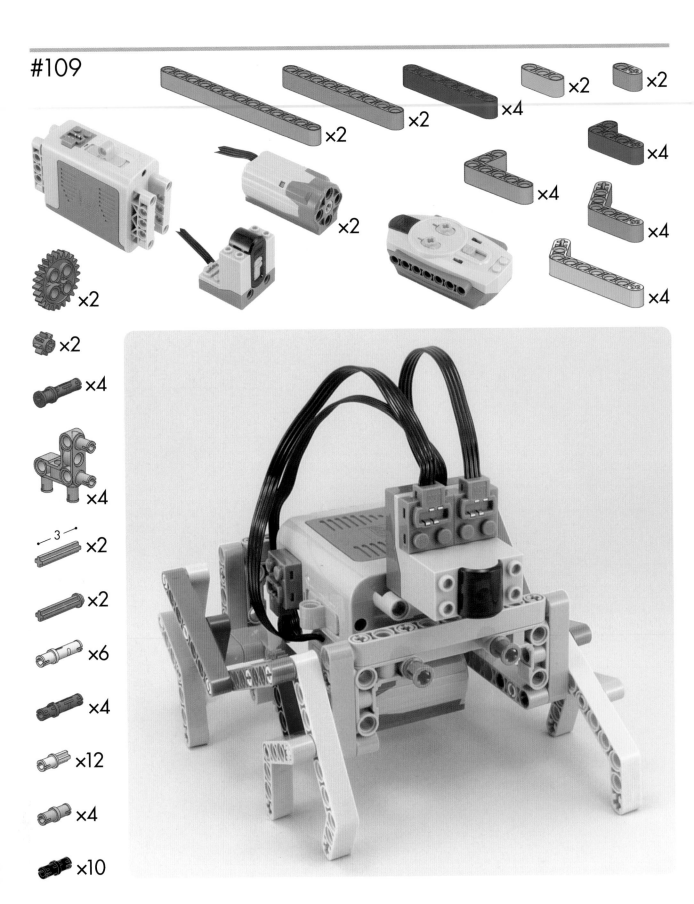

×2 ×2 ×4 ×2 ×2

×2

×4 ×4

×4 ×4

×4

×2

×4

×4

—3—×2

×2

×6

×4

×12

×4

×10

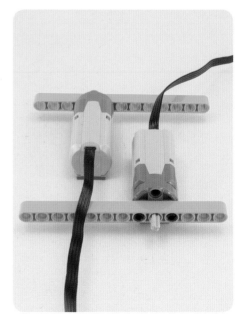

Moving like an inchworm

#110

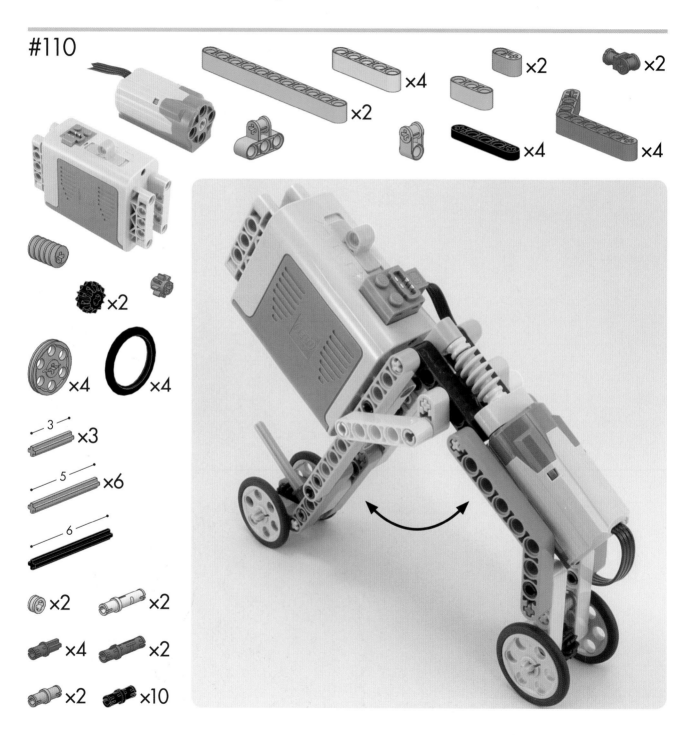

×2 ×4 ×2 ×2

×2 ×4 ×4

×2

×4 ×4

×3
3

×6
5

6

×2 ×2

×4 ×2

×2 ×10

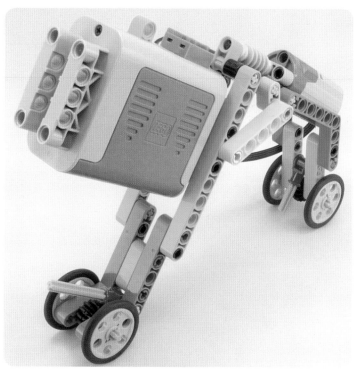

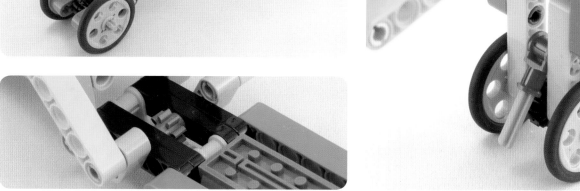

#111

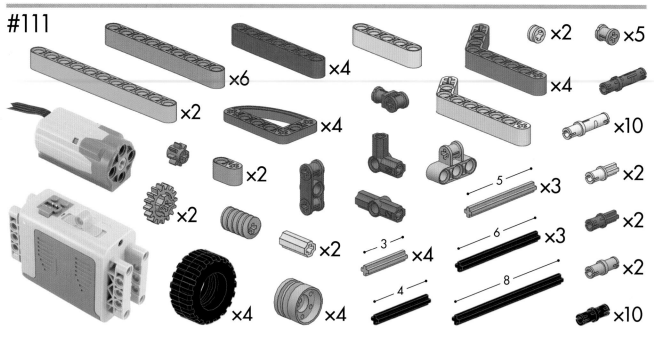

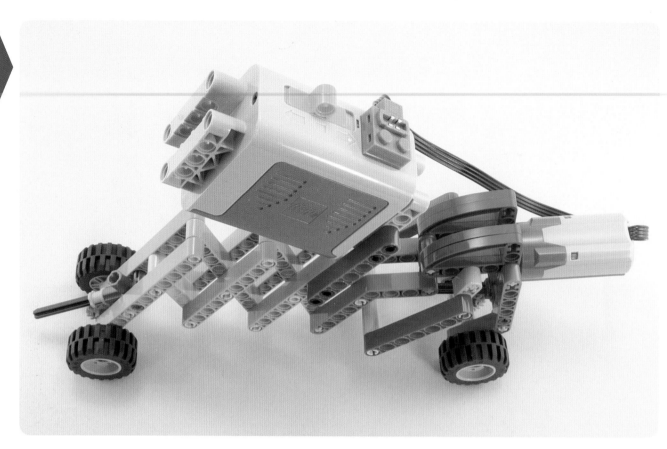

Moving through vibration

#112

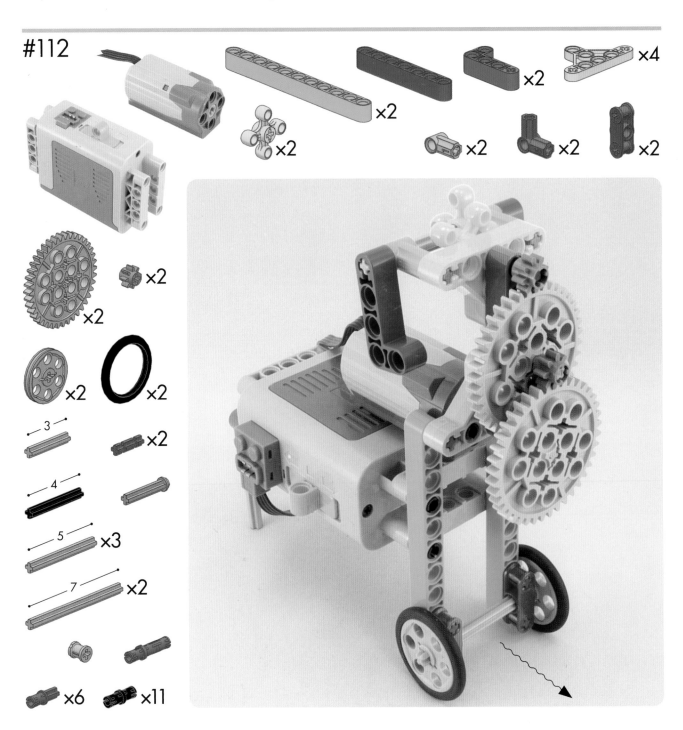

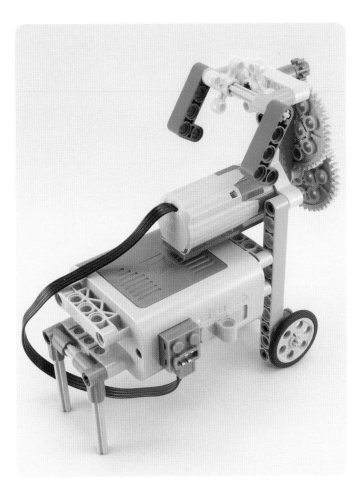

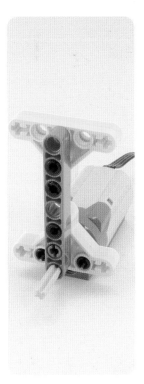

#113

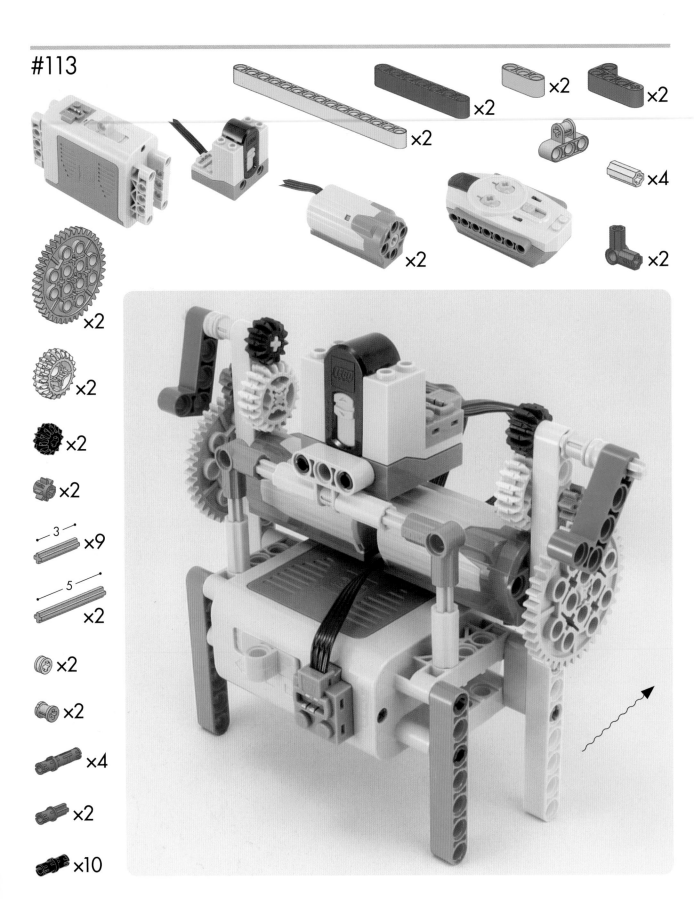

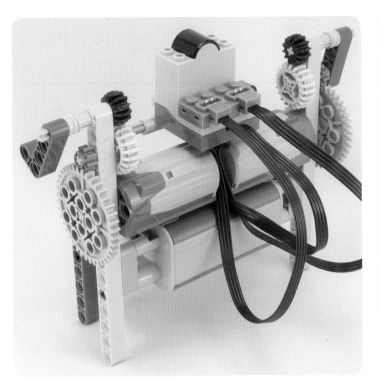

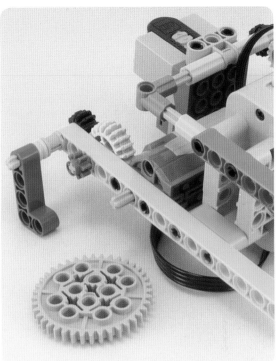

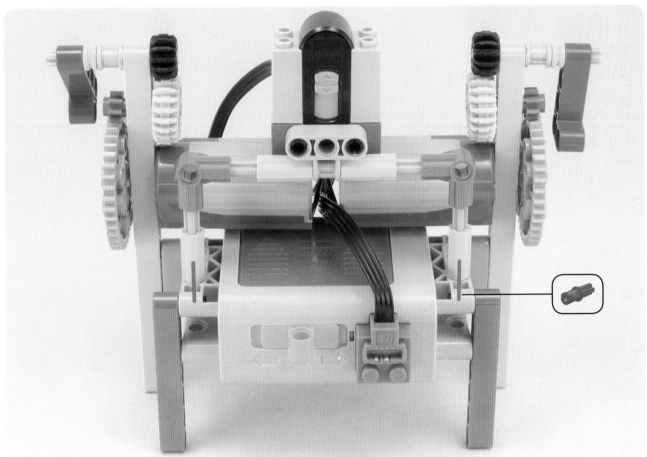

Moving in other ways

#114

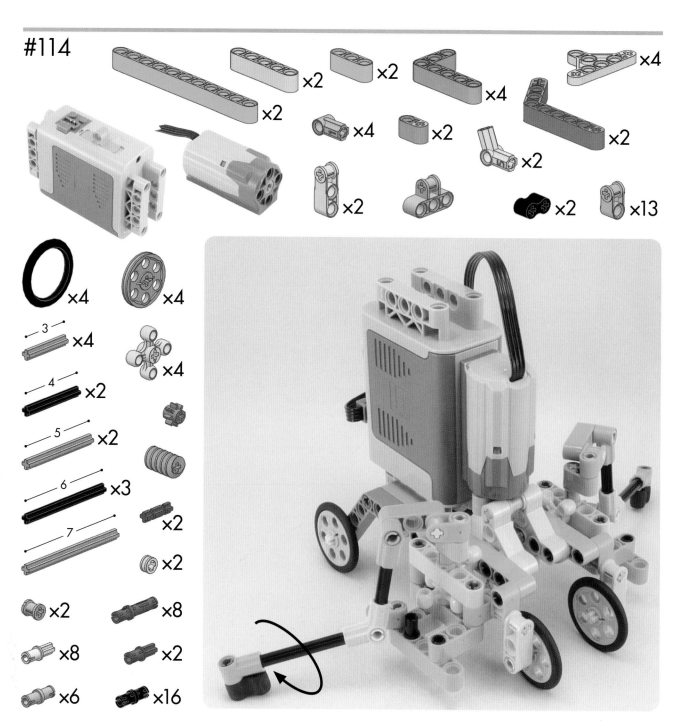

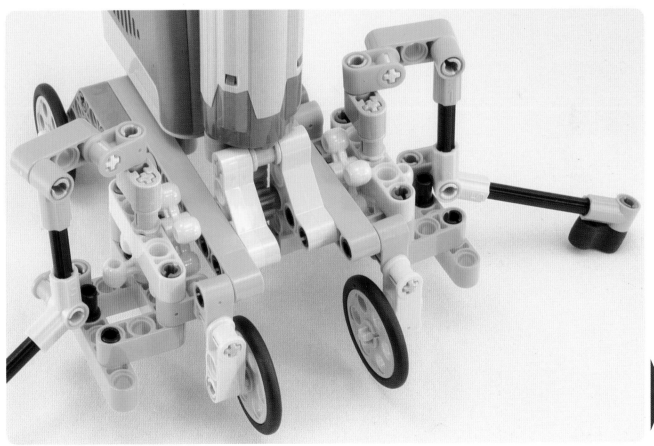

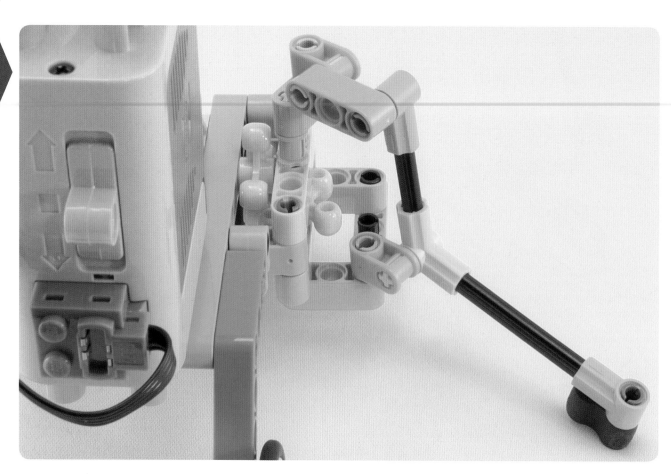

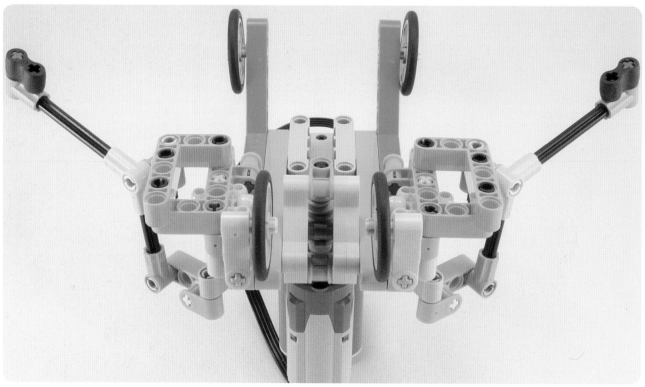

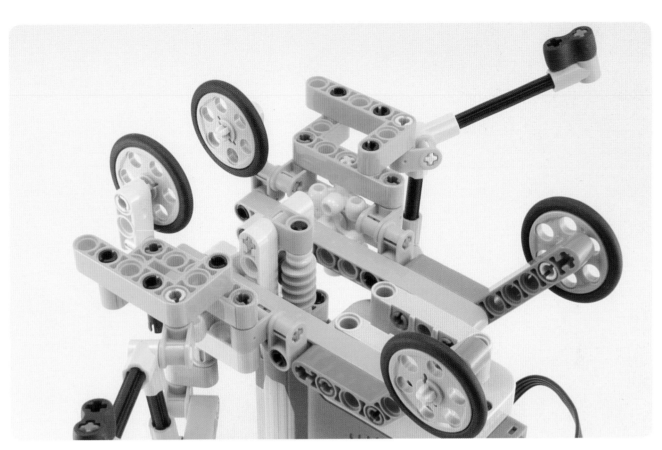

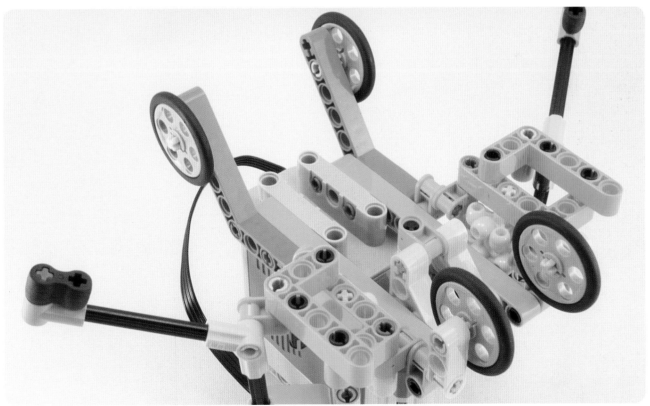

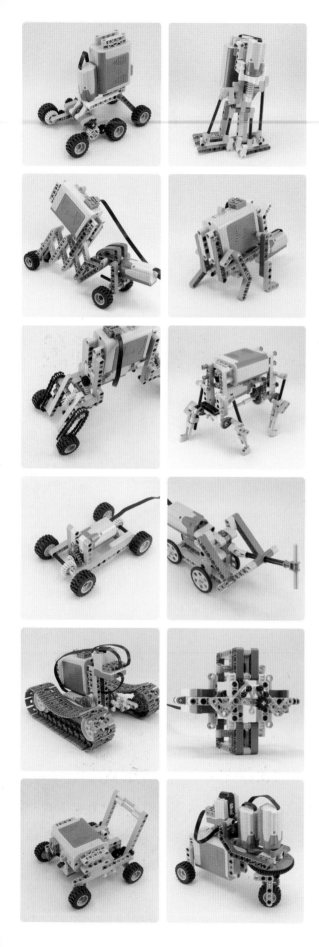

PART 3

Special Mechanisms

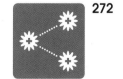

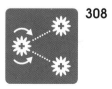

Intermittent motion

#115

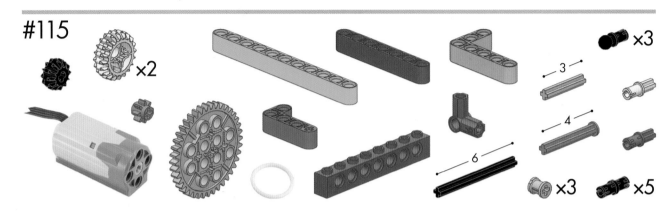

×2

×3

3

4

6

×3

×5

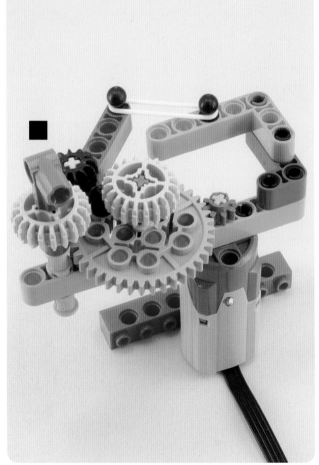

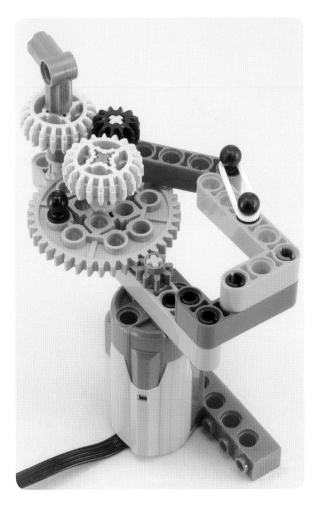

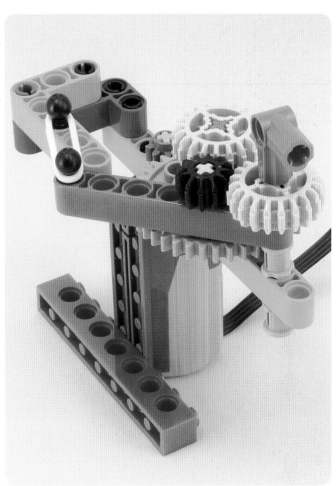

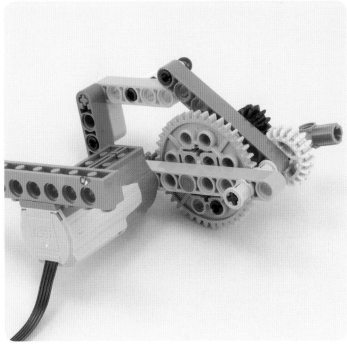

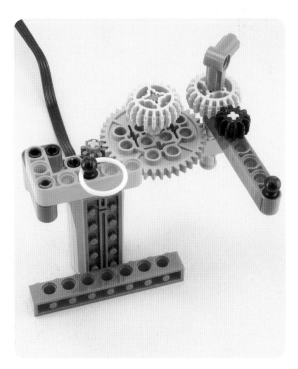

#116

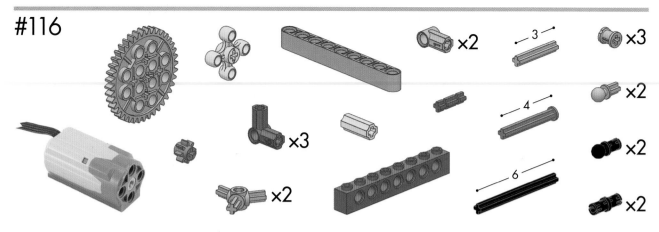

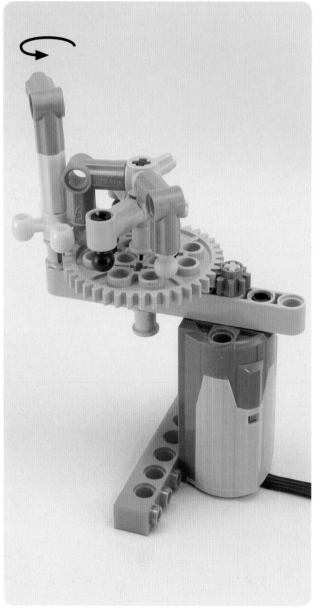

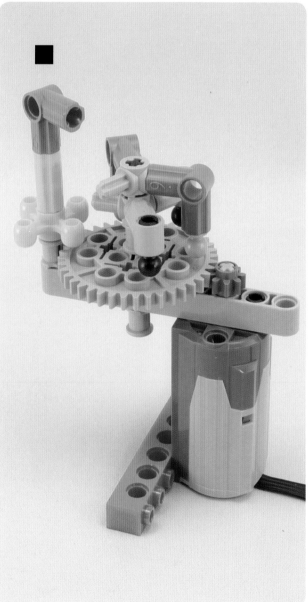

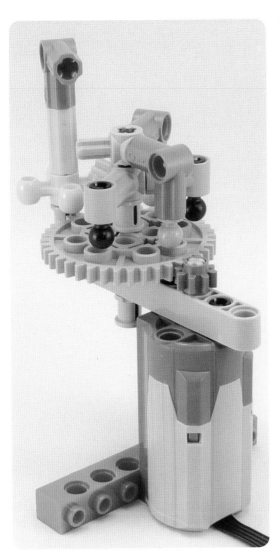

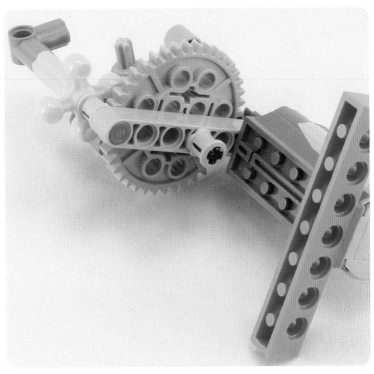

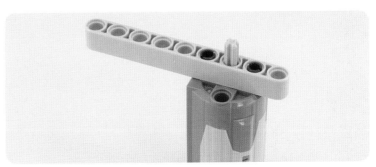

#117

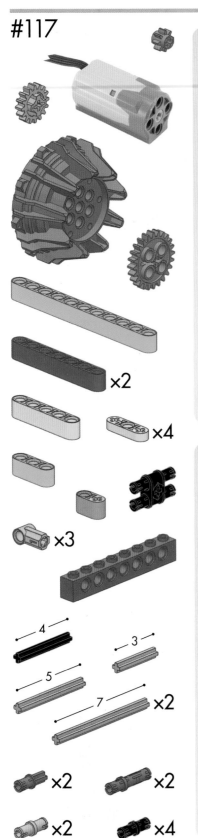

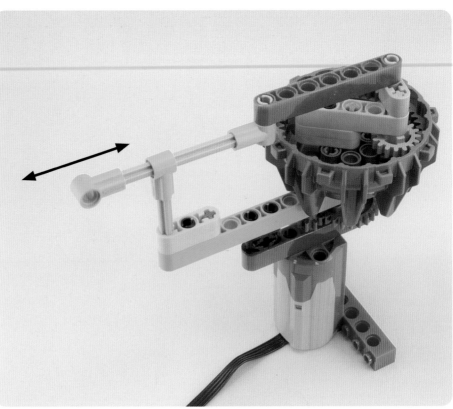

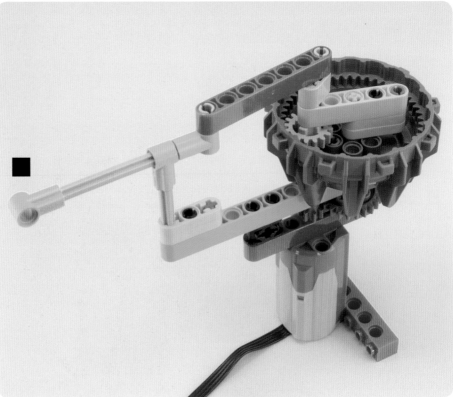

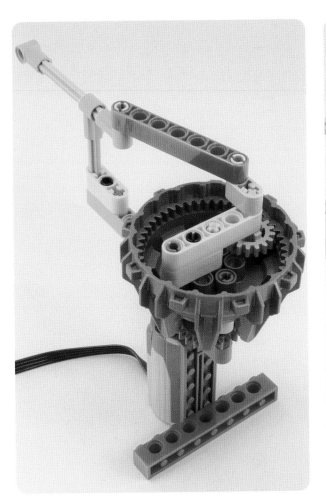

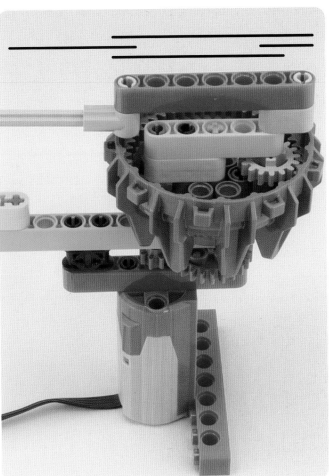

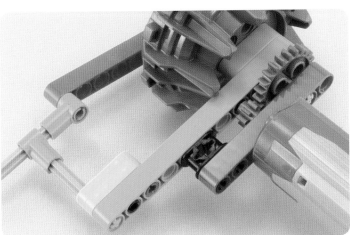

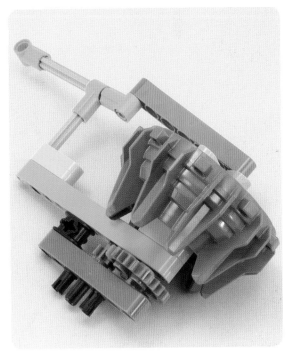

Smooth changes of rotation speed

#118

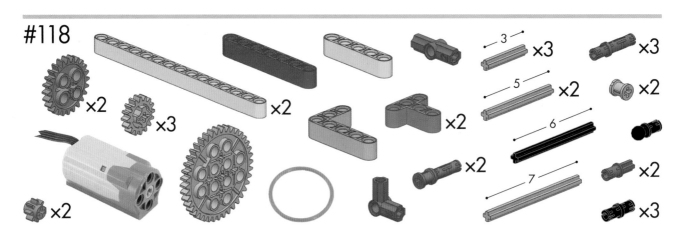

×2   ×3   ×2   ×2   ×3   ×3   ×2   ×2   ×2   3   5   6   7

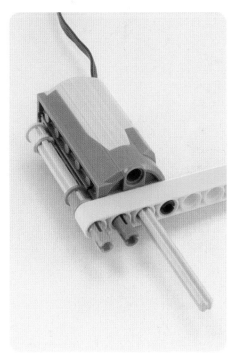

#119

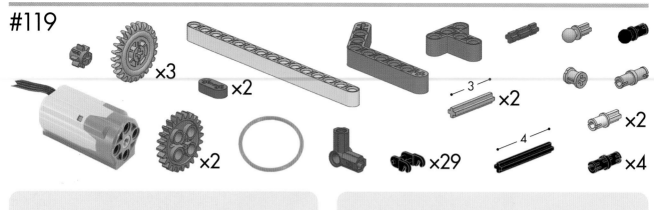

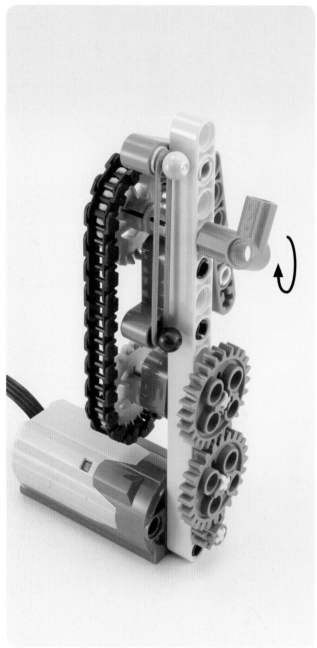

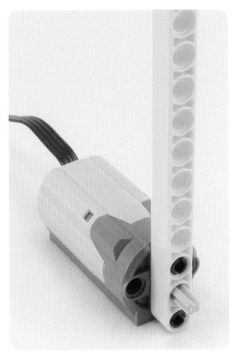

#120

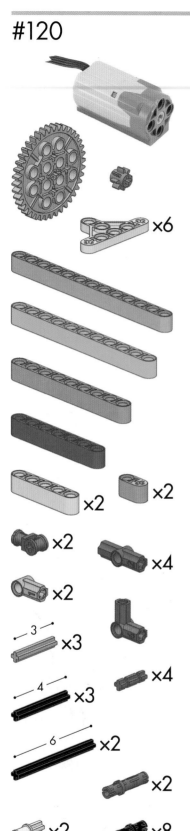

×6

×2

×2

×2

×4

×2

3 ×3

4 ×3

6 ×2

×2

×2

×8

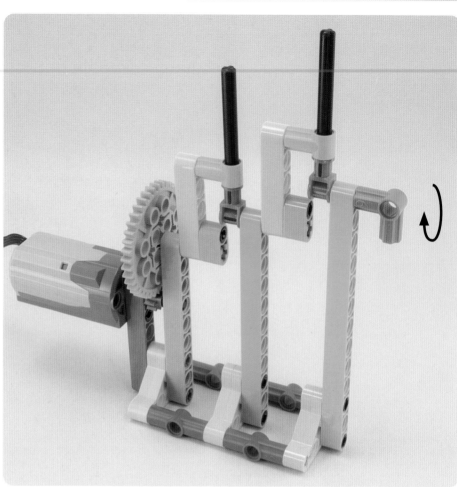

#121

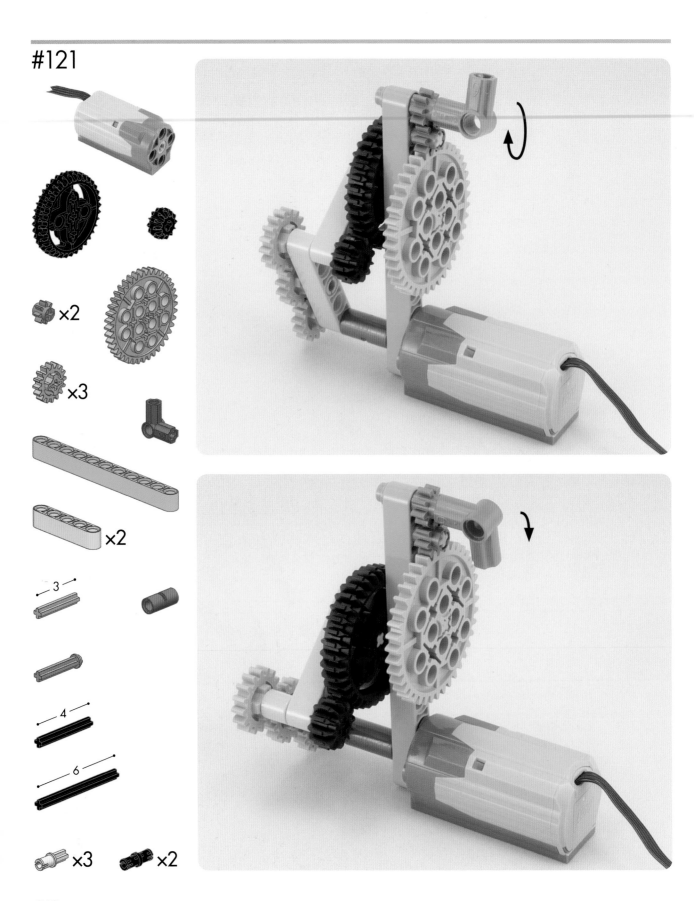

×2

×3

×2

3

4

6

×3 ×2

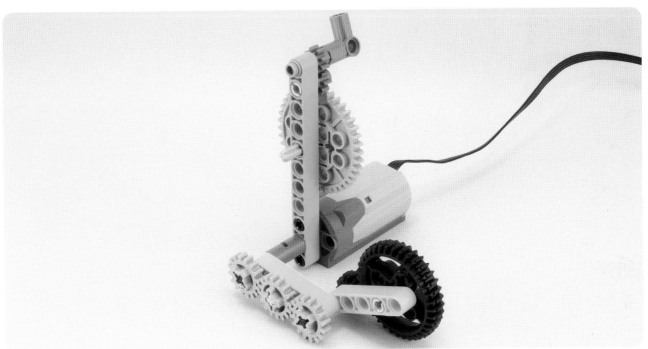

Switching rotational direction

#122

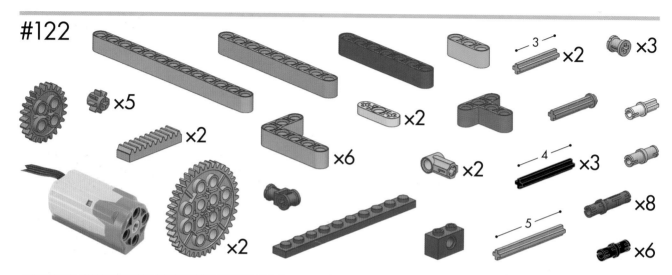

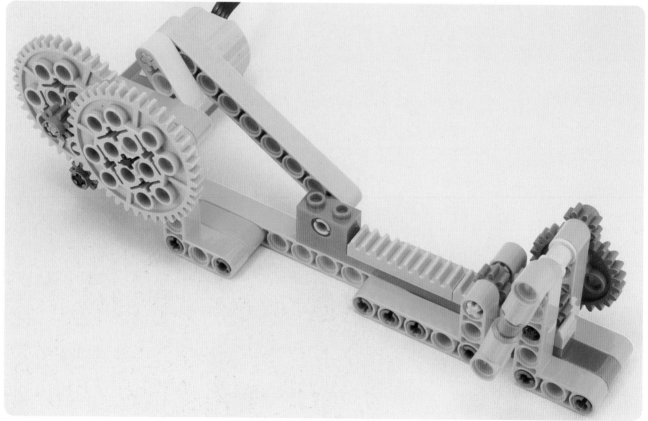

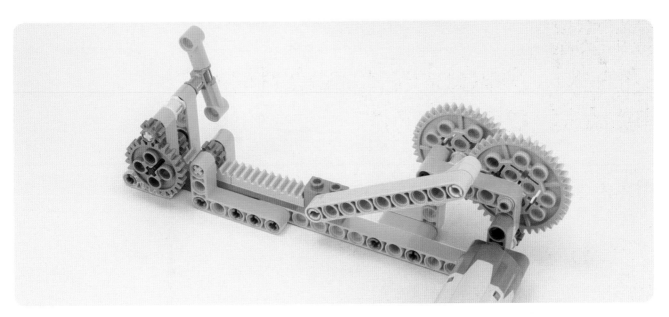

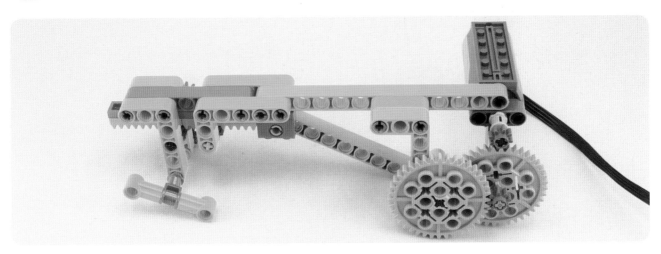

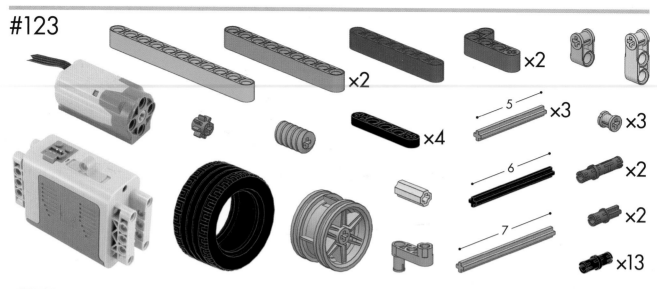

123

×2 ×2

×4 5 ×3 ×3

6 ×2

×2

7 ×13

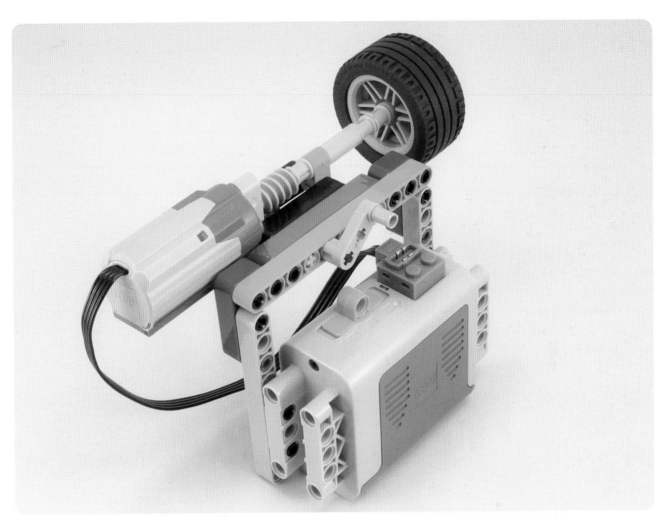

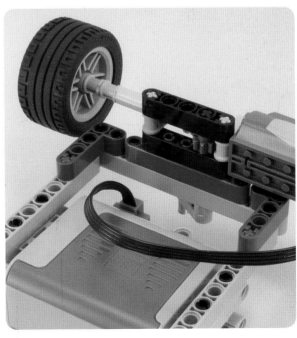

#124

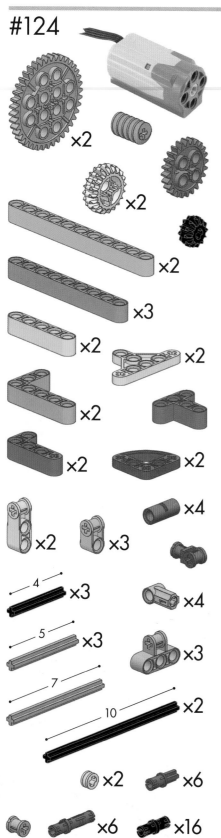

×2

×2

×2

×3

×2

×2

×2

×2

×2

×2

×2

×3

×4

4 ×3

×4

5 ×3

×3

7

10 ×2

×2

×6

×6

×16

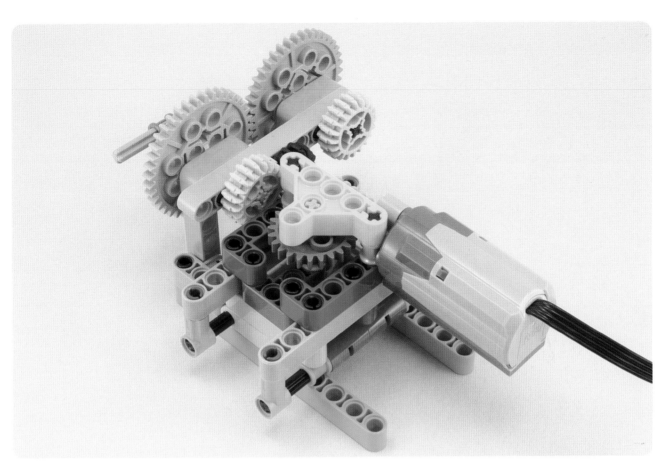

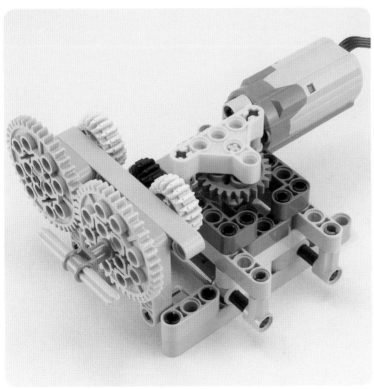

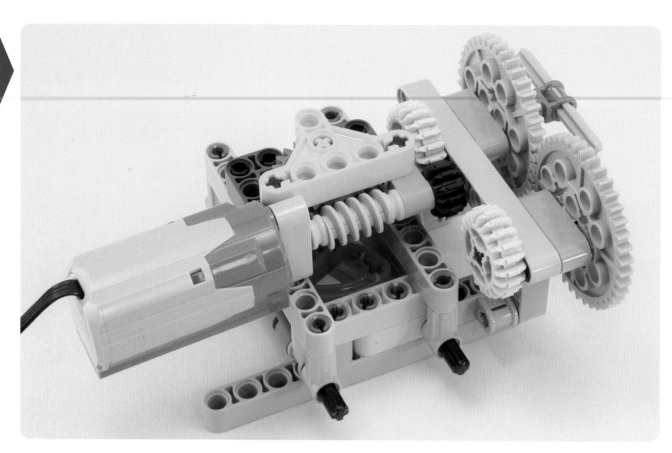

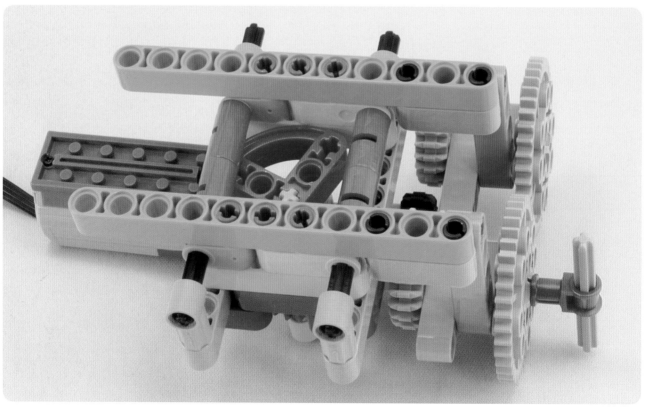

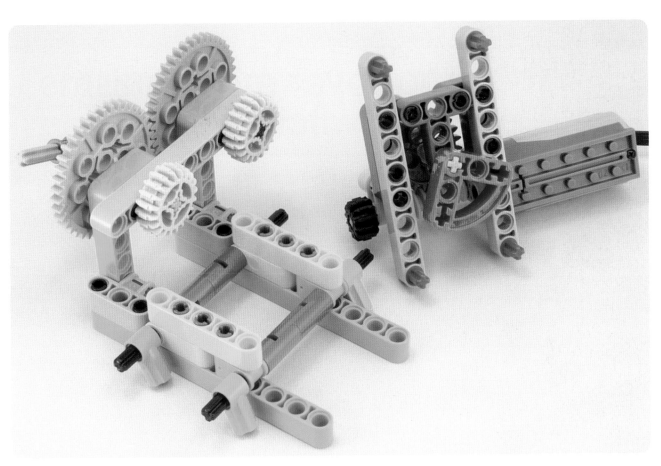

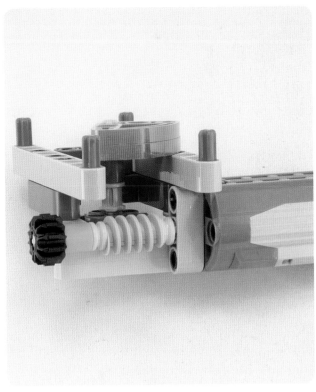

Changeover mechanisms using a switch

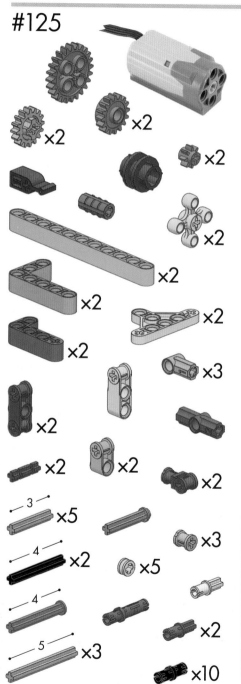

#125

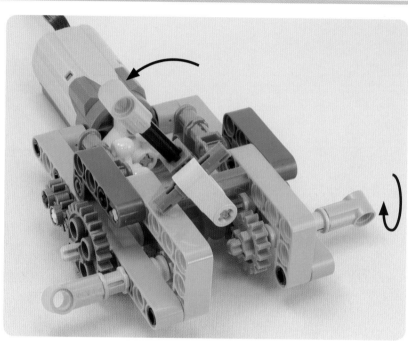

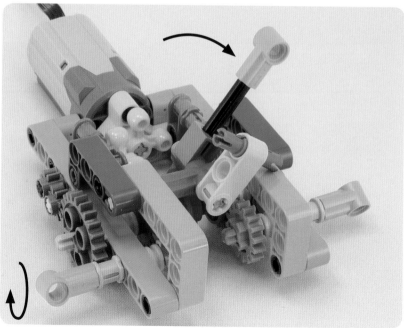

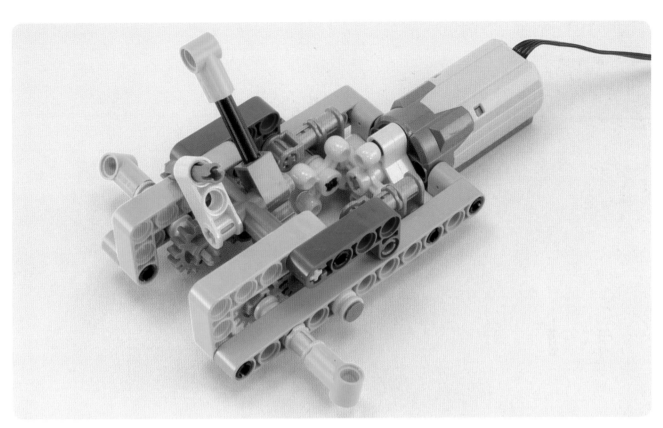

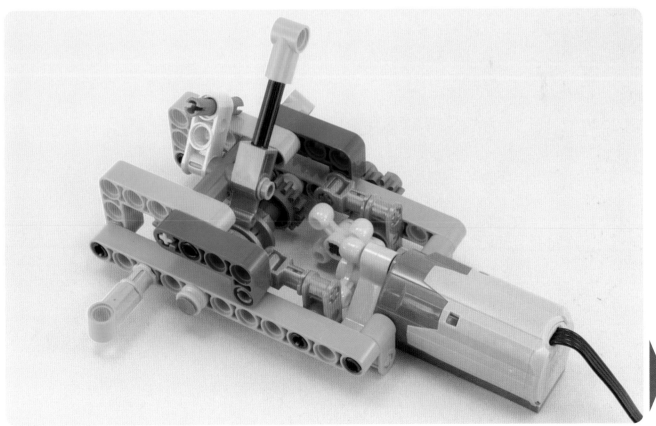

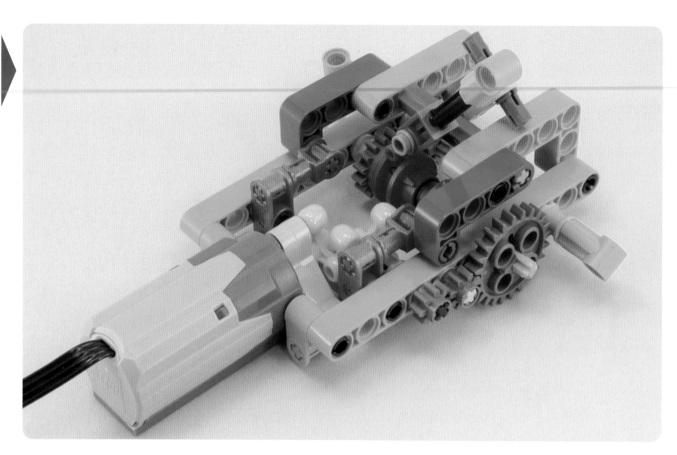

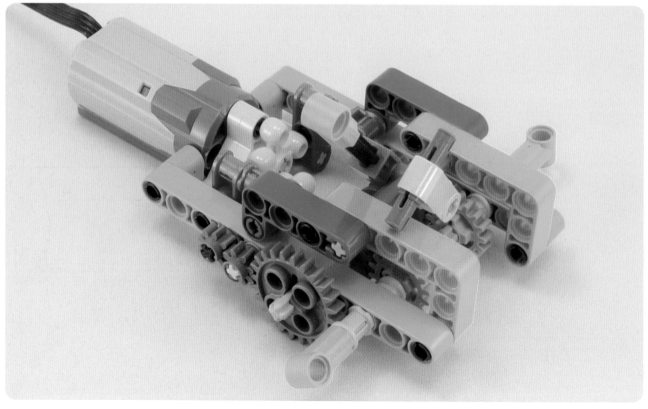

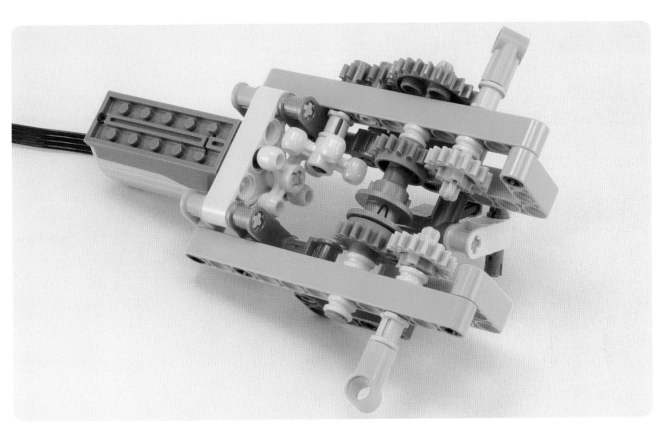

#126

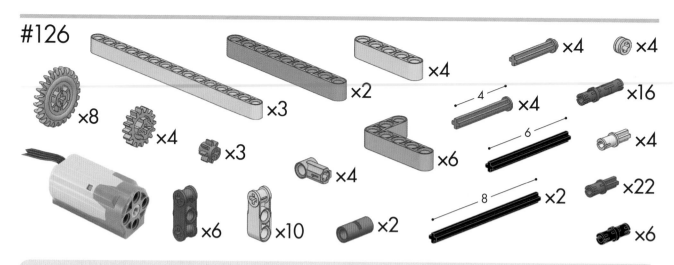

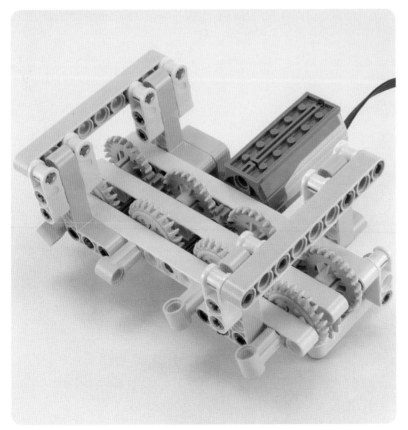

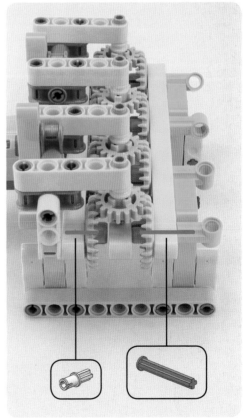

#127

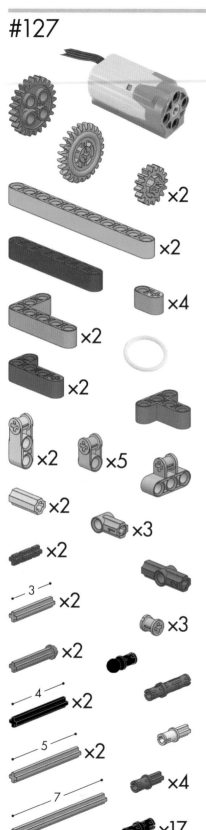

×2
×2
×4
×2
×2
×2
×5
×2
×3
×2
×2
3 ×2
×3
×2
4 ×2
5 ×2
×4
7
×17

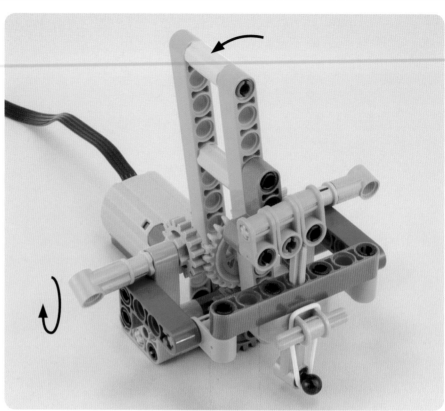

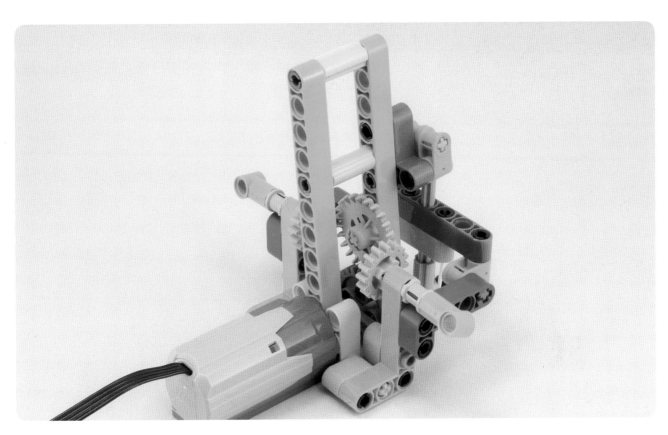

#128

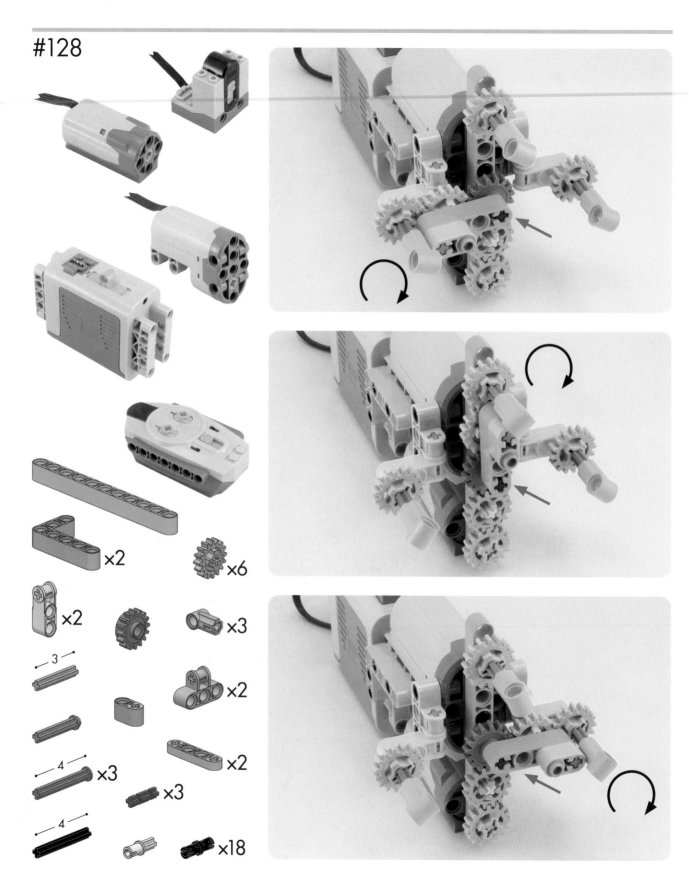

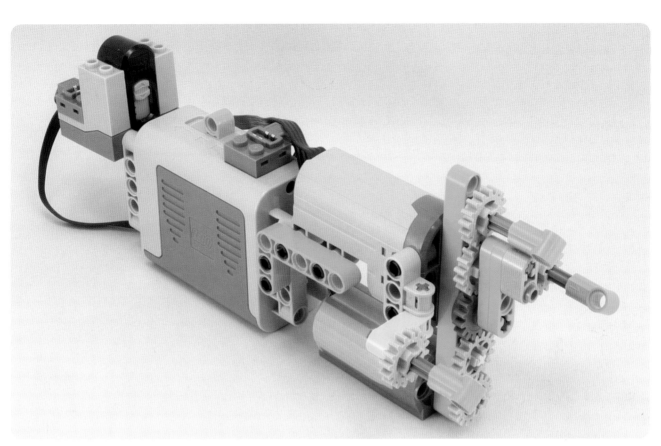

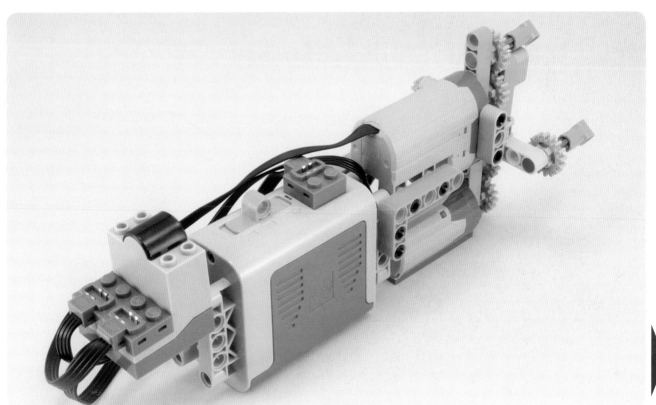

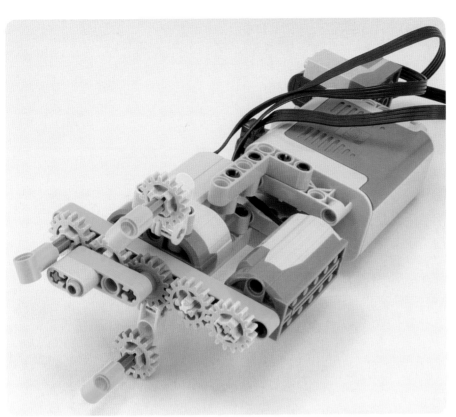

#129

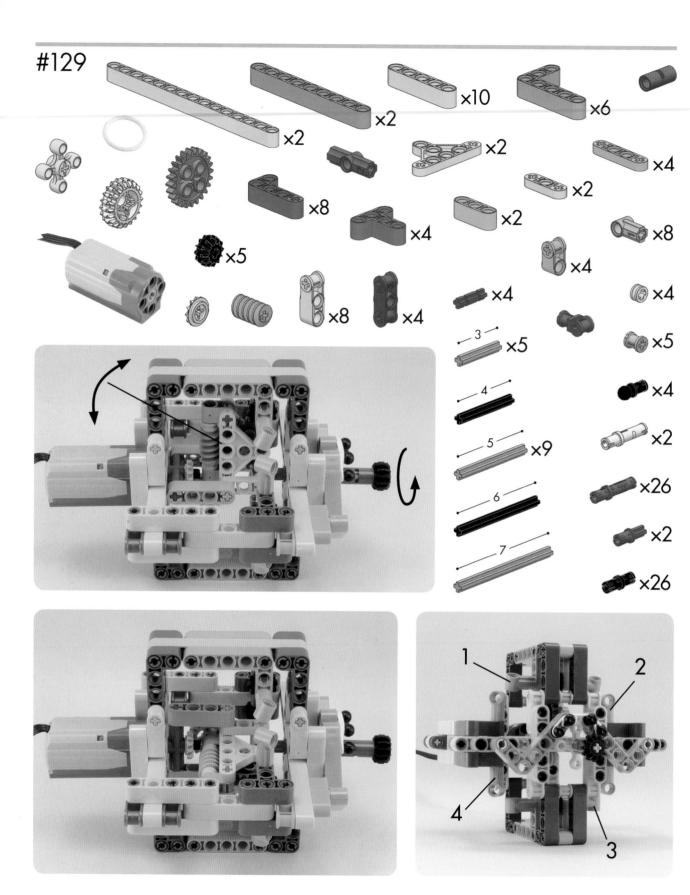

×2
×2
×10
×6
×2
×4
×8
×2
×2
×4
×4
×4
×4
×5
×8
×4
×8

3 ×5
4
5 ×9
6
7

×4
×4
×5
×4
×2
×26
×2
×26

1
2
3
4

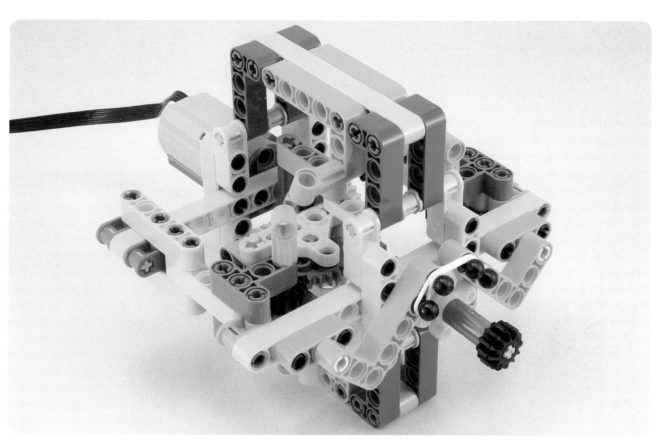

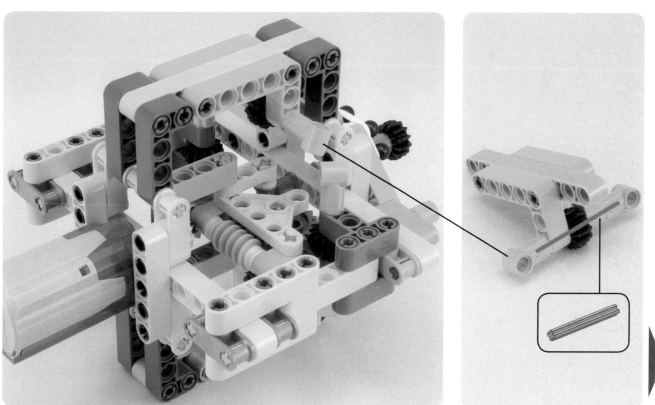

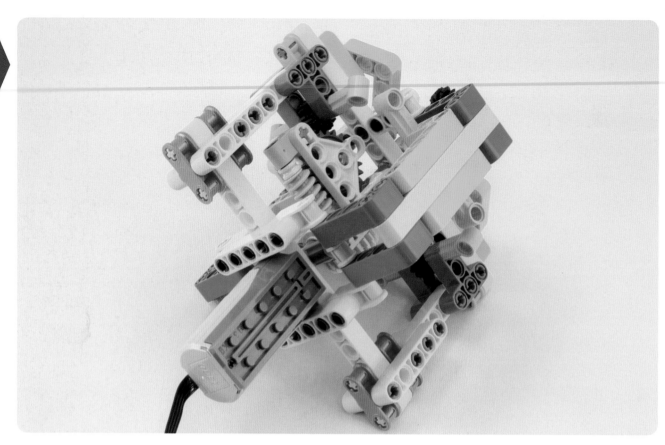

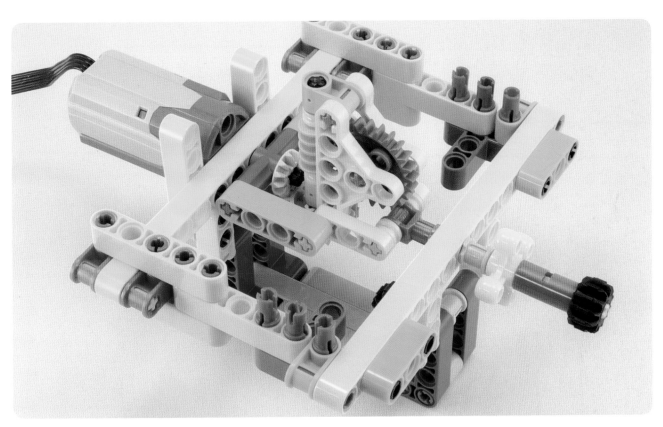

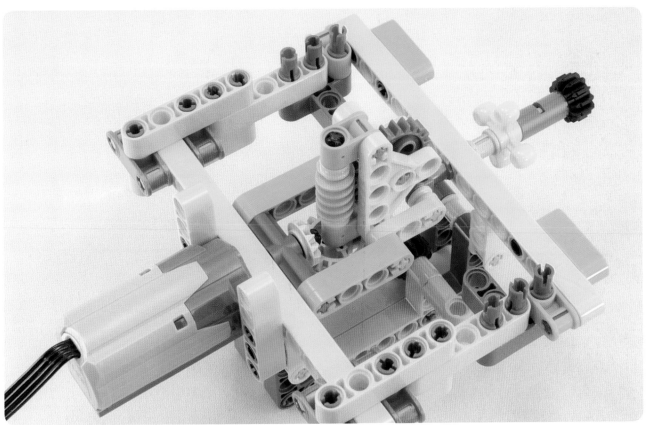

Transmissions

#130

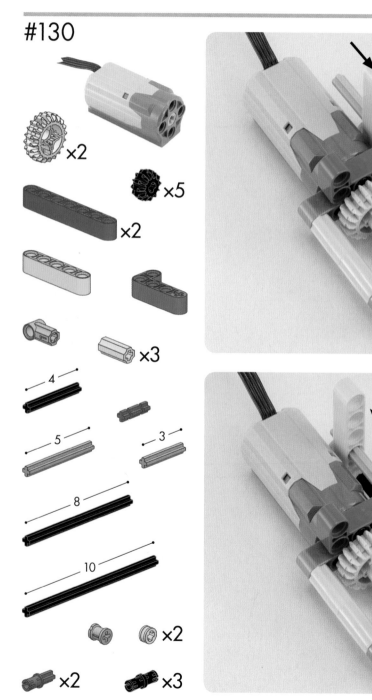

×2

×5

×2

×3

4

5

3

8

10

×2

×2

×2

×3

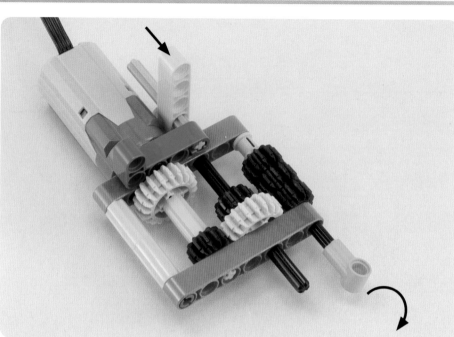

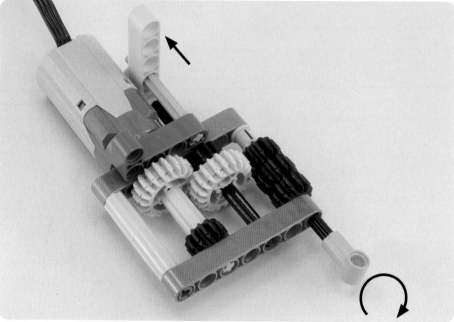

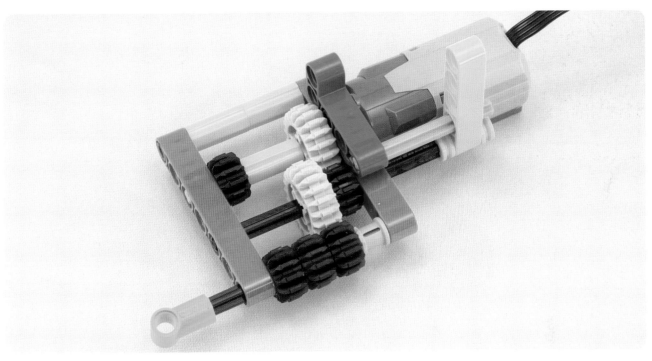

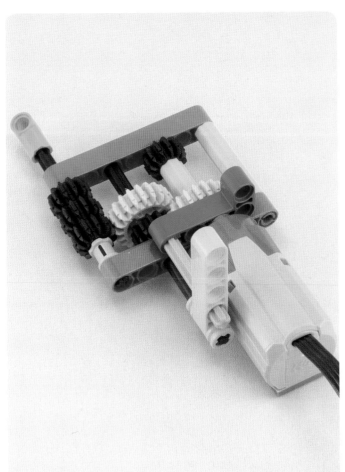

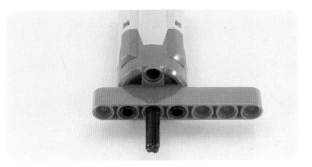

#131

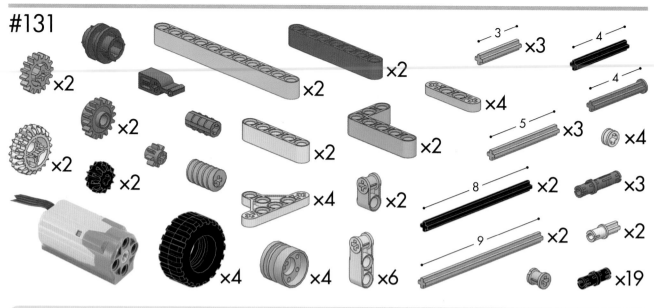

#132

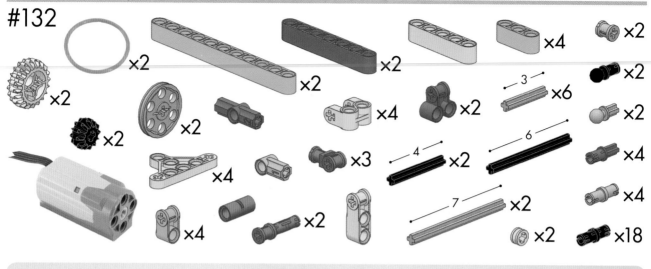

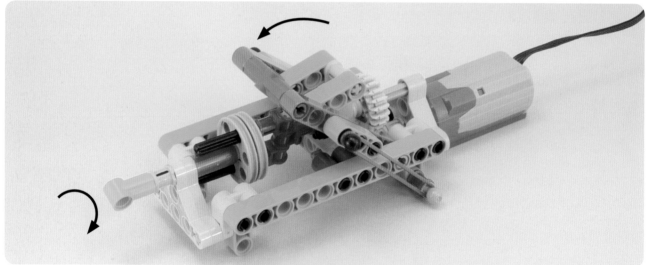

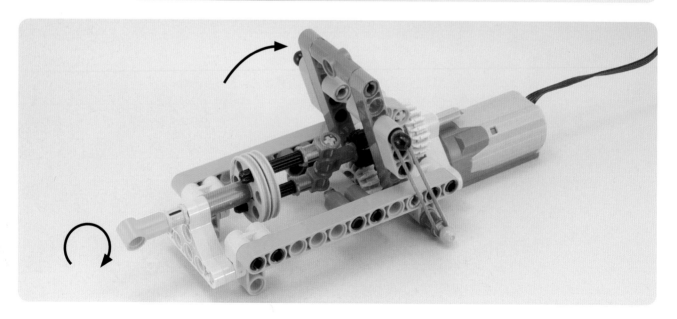

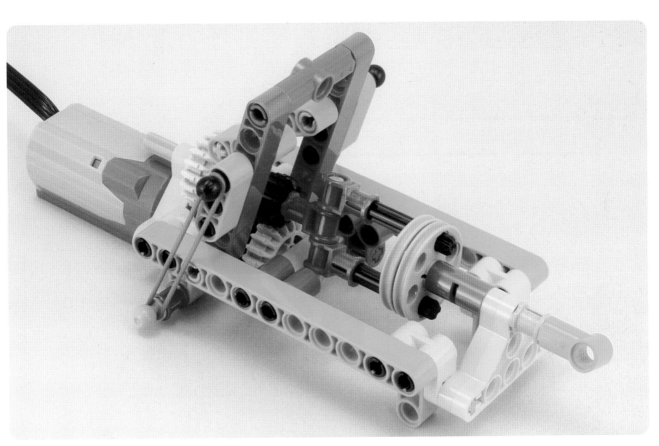

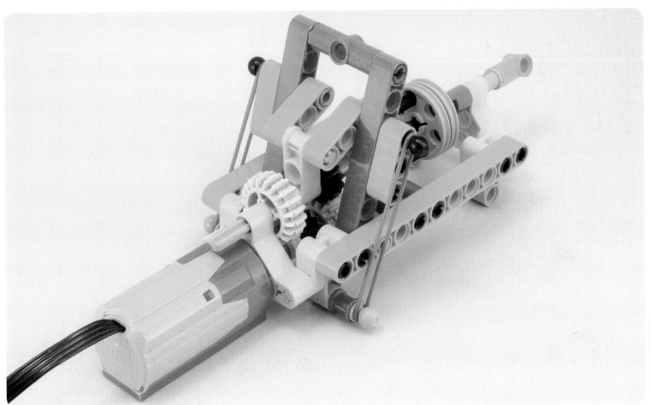

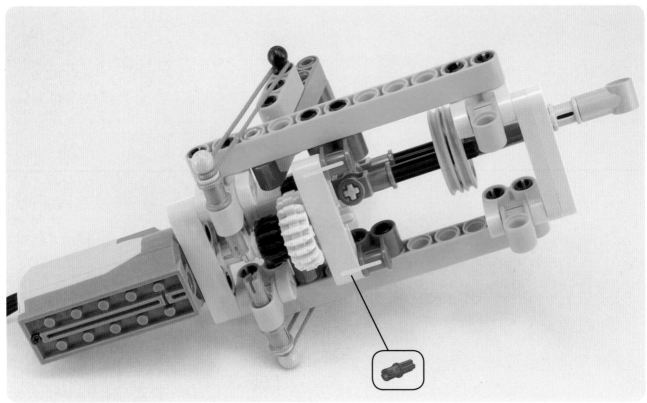

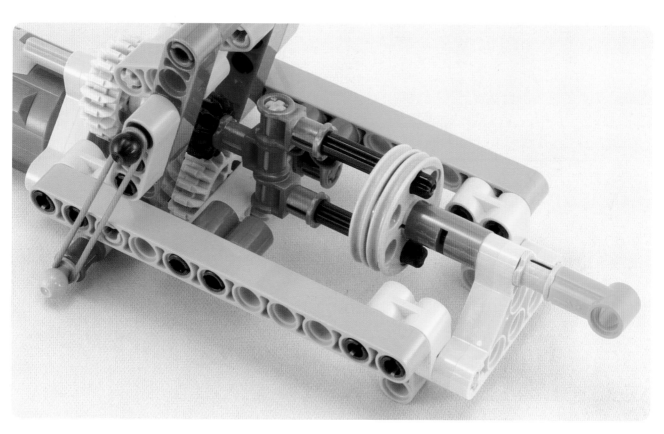

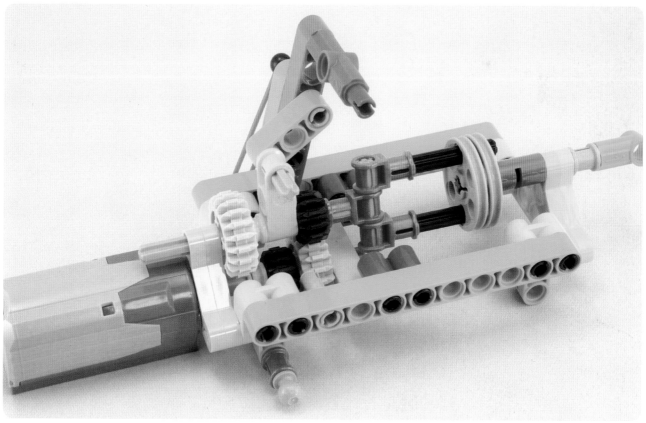

#133

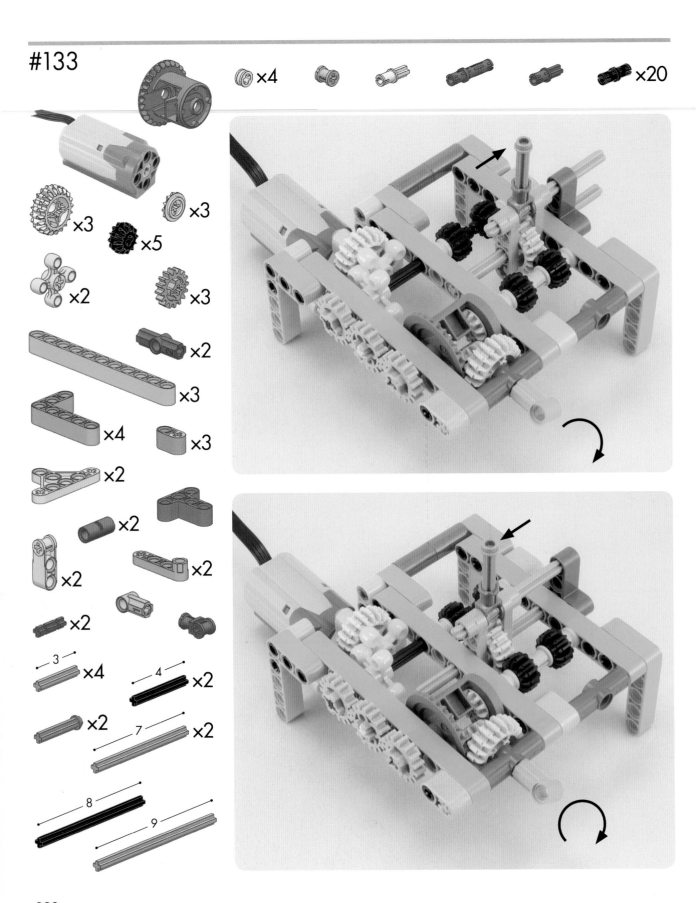

×4

×20

×3

×3

×5

×2

×3

×3

×4

×3

×2

×2

×2

×2

×2

3 ×4

4 ×2

×2

7 ×2

8

9

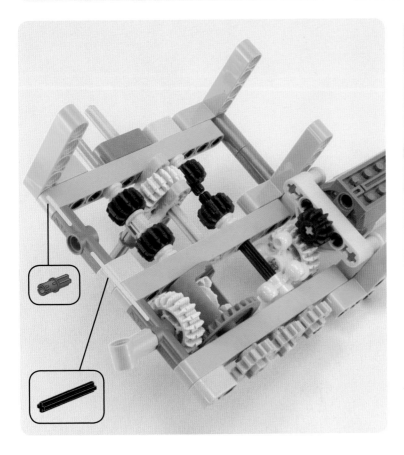

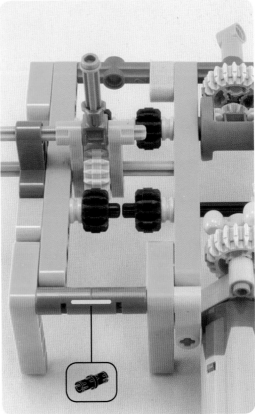

#134

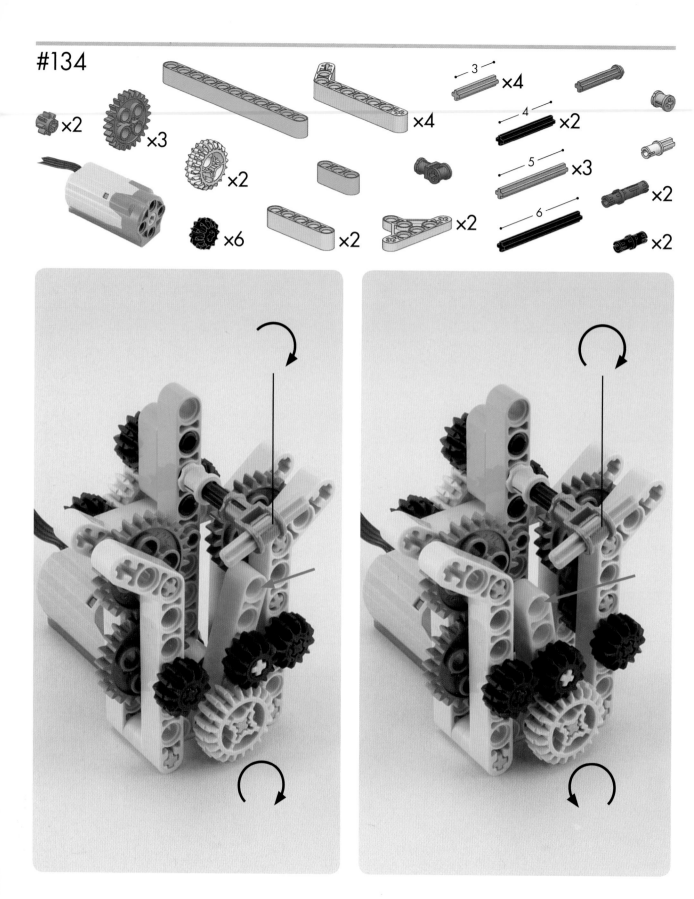

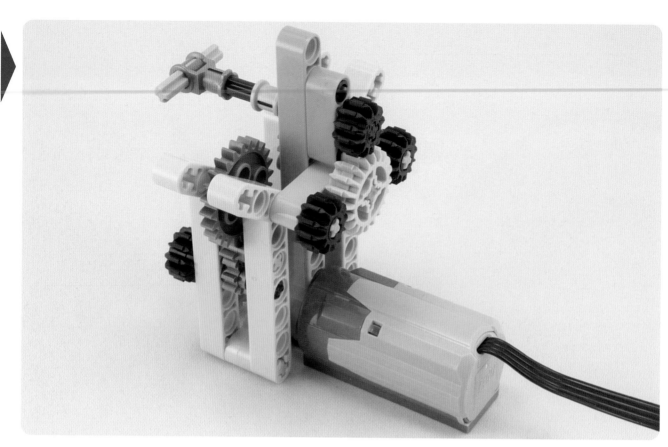

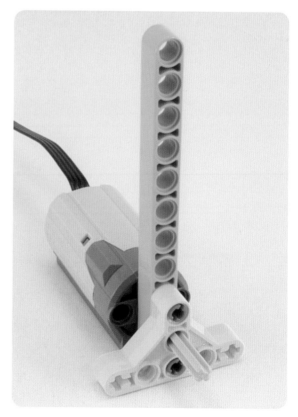

#135

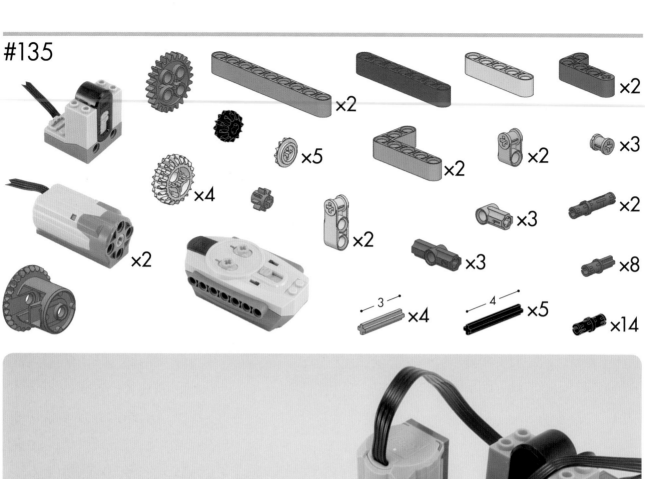

×2 ×2 ×5 ×4 ×2 ×3 ×2 ×3 ×2 ×2 ×3 ×8 ×14

Eight speeds

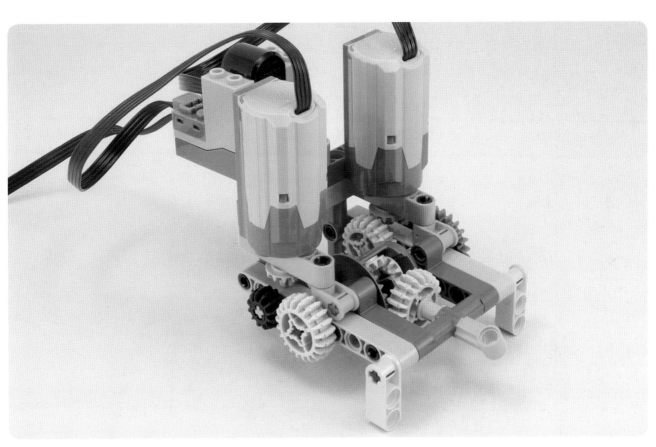

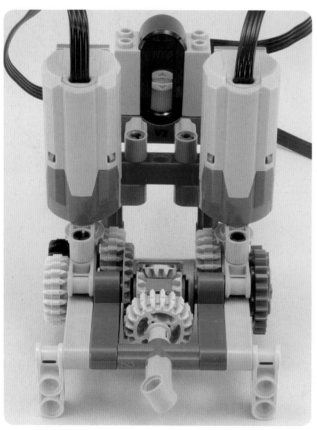

Changeover mechanisms using rotational direction

#136

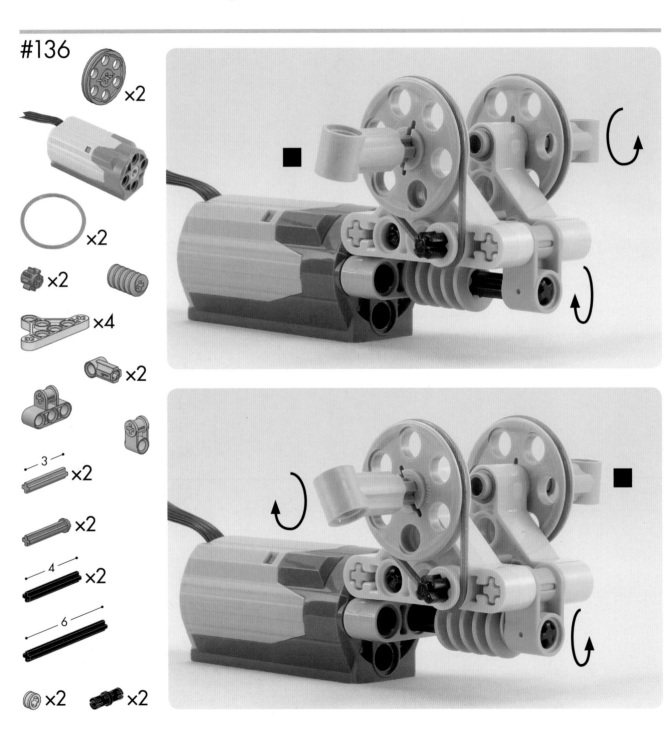

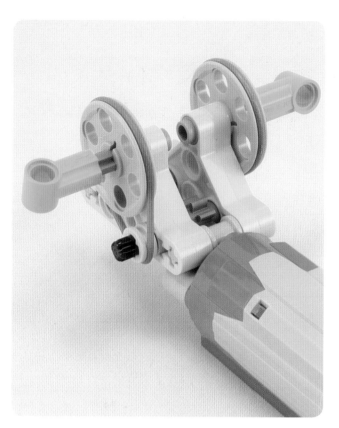

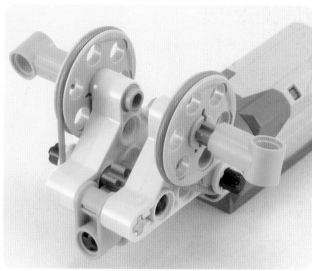

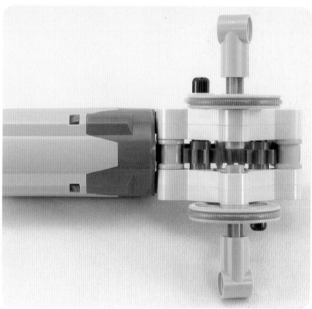

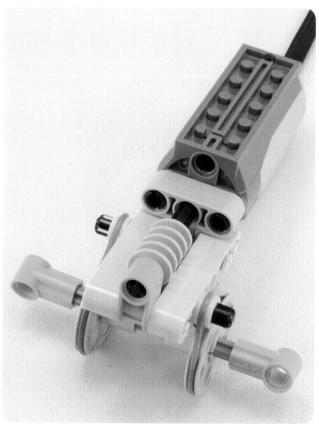

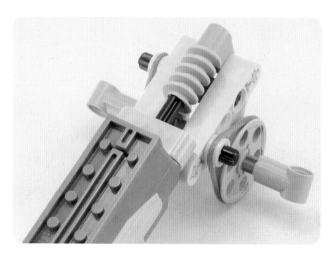

137

×3 ×2 ×2

×3 ×2

×3 ×2 ×4

4 4

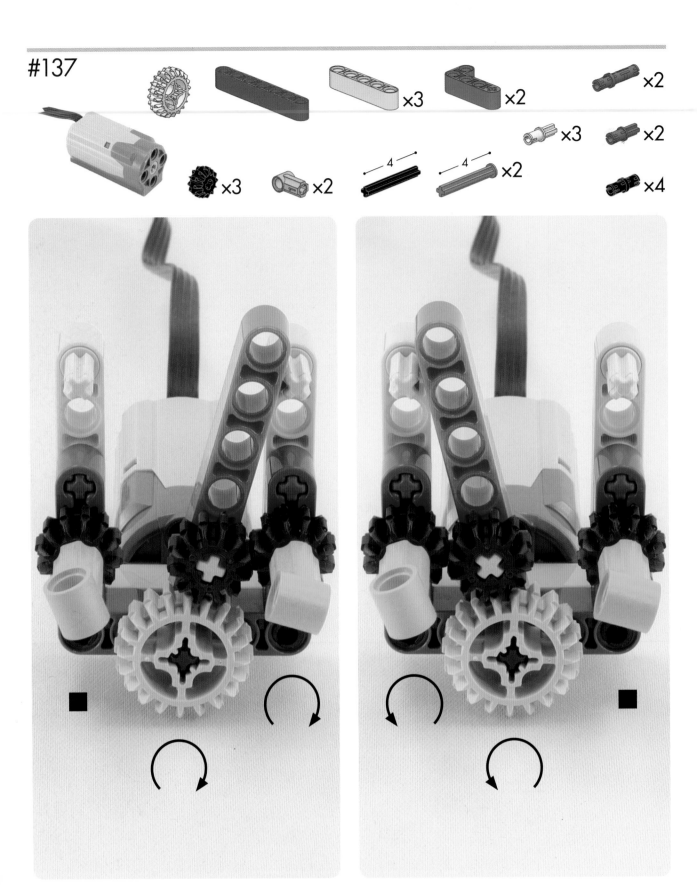

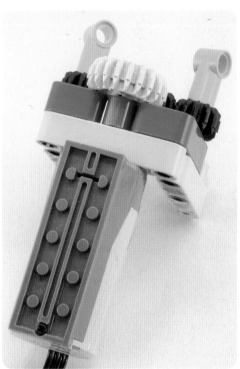

#138

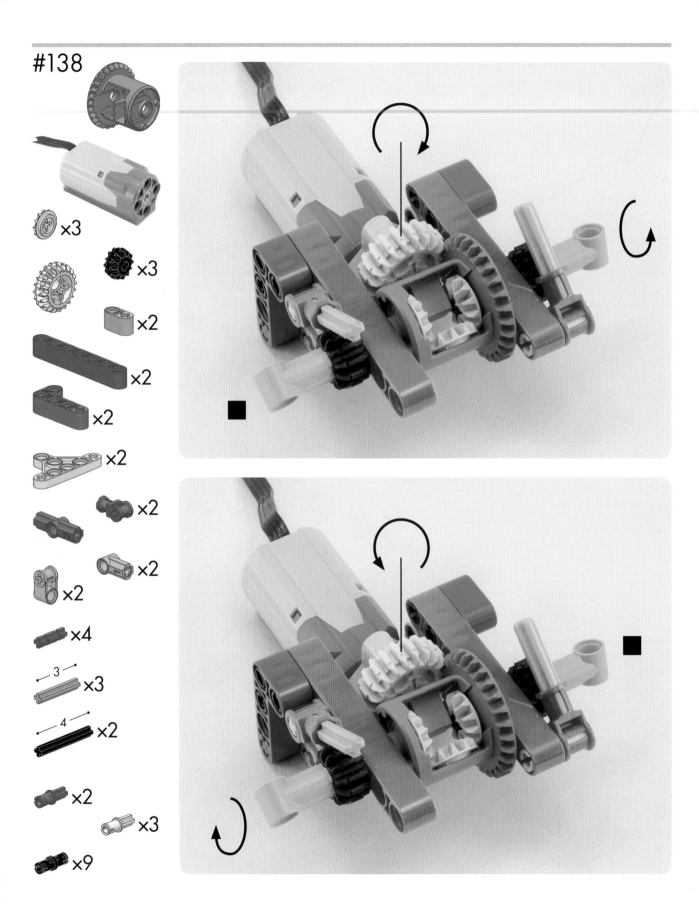

Parts list

This is the maximum number of this part needed to build any single model in this book.

This is the maximum number of this part needed to build any single model in *both* volumes of *The LEGO Power Functions Idea Book* (*Machines and Mechanisms* and *Cars and Contraptions*).

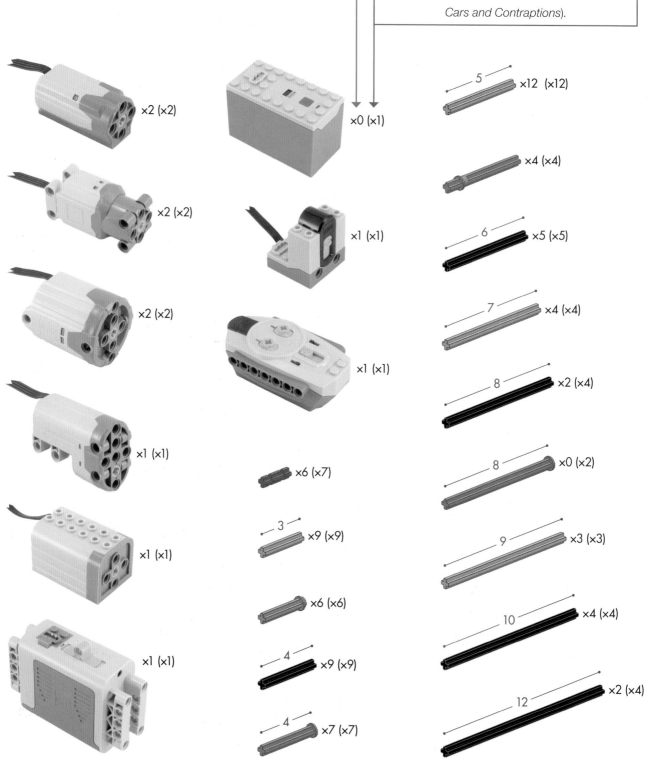

×2 (×2)

×2 (×2)

×2 (×2)

×1 (×1)

×1 (×1)

×1 (×1)

×0 (×1)

×1 (×1)

×1 (×1)

×6 (×7)

3 ×9 (×9)

×6 (×6)

4 ×9 (×9)

4 ×7 (×7)

5 ×12 (×12)

×4 (×4)

6 ×5 (×5)

7 ×4 (×4)

8 ×2 (×4)

8 ×0 (×2)

9 ×3 (×3)

10 ×4 (×4)

12 ×2 (×4)

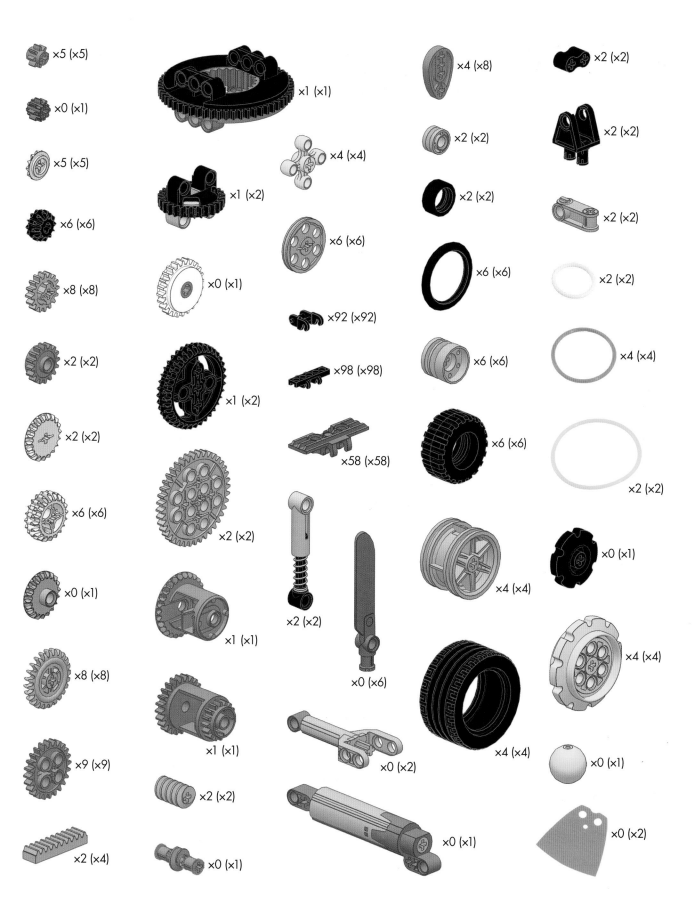

×5 (×5)

×0 (×1)

×5 (×5)

×6 (×6)

×8 (×8)

×2 (×2)

×2 (×2)

×6 (×6)

×0 (×1)

×8 (×8)

×9 (×9)

×2 (×4)

×1 (×1)

×1 (×2)

×0 (×1)

×1 (×2)

×2 (×2)

×1 (×1)

×1 (×1)

×2 (×2)

×0 (×1)

×4 (×4)

×6 (×6)

×92 (×92)

×98 (×98)

×58 (×58)

×2 (×2)

×0 (×6)

×0 (×2)

×0 (×1)

×4 (×8)

×2 (×2)

×2 (×2)

×6 (×6)

×6 (×6)

×6 (×6)

×4 (×4)

×4 (×4)

×2 (×2)

×2 (×2)

×2 (×2)

×2 (×2)

×4 (×4)

×2 (×2)

×0 (×1)

×4 (×4)

×0 (×1)

×0 (×2)

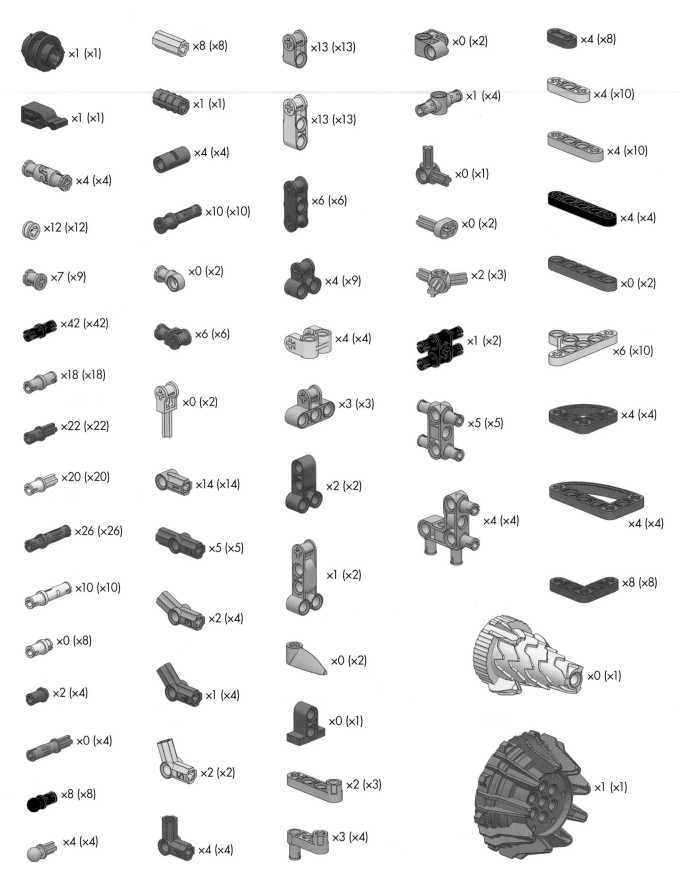

×1 (×1)

×8 (×8)

×13 (×13)

×0 (×2)

×4 (×8)

×1 (×1)

×1 (×1)

×13 (×13)

×1 (×4)

×4 (×10)

×4 (×4)

×4 (×4)

×0 (×1)

×4 (×10)

×12 (×12)

×10 (×10)

×6 (×6)

×0 (×2)

×4 (×4)

×7 (×9)

×0 (×2)

×4 (×9)

×2 (×3)

×0 (×2)

×42 (×42)

×6 (×6)

×4 (×4)

×1 (×2)

×6 (×10)

×18 (×18)

×0 (×2)

×3 (×3)

×5 (×5)

×4 (×4)

×22 (×22)

×14 (×14)

×2 (×2)

×4 (×4)

×4 (×4)

×20 (×20)

×5 (×5)

×1 (×2)

×26 (×26)

×2 (×4)

×0 (×2)

×8 (×8)

×10 (×10)

×1 (×4)

×0 (×1)

×0 (×1)

×0 (×8)

×2 (×4)

×2 (×2)

×0 (×4)

×2 (×3)

×8 (×8)

×4 (×4)

×4 (×4)

×3 (×4)

×1 (×1)

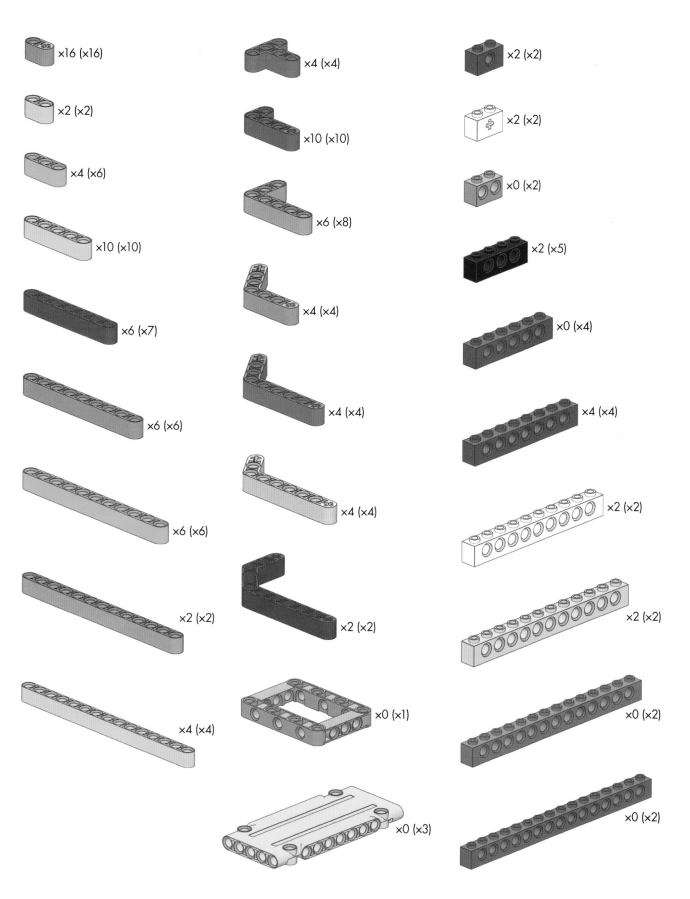

×16 (×16)

×2 (×2)

×4 (×6)

×10 (×10)

×6 (×7)

×6 (×6)

×6 (×6)

×2 (×2)

×4 (×4)

×4 (×4)

×10 (×10)

×6 (×8)

×4 (×4)

×4 (×4)

×4 (×4)

×2 (×2)

×0 (×1)

×0 (×3)

×2 (×2)

×2 (×2)

×0 (×2)

×2 (×5)

×0 (×4)

×4 (×4)

×2 (×2)

×2 (×2)

×0 (×2)

×0 (×2)

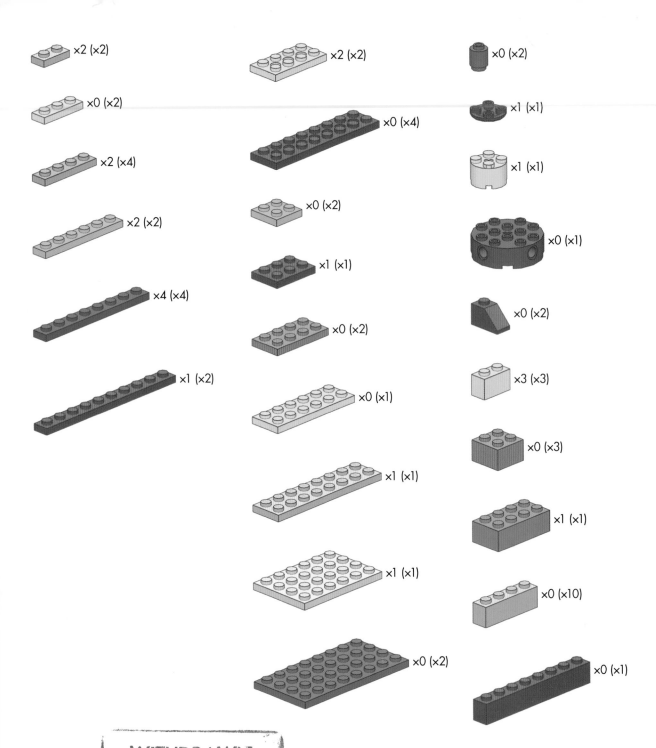

×2 (×2)

×2 (×2)

×0 (×2)

×0 (×2)

×1 (×1)

×2 (×4)

×0 (×4)

×1 (×1)

×2 (×2)

×0 (×2)

×0 (×1)

×4 (×4)

×1 (×1)

×0 (×2)

×1 (×2)

×0 (×2)

×3 (×3)

×0 (×1)

×0 (×3)

×1 (×1)

×1 (×1)

×1 (×1)

×0 (×10)

×0 (×2)

×0 (×1)